Lecture Notes in Physics

Bisher erschienen/Already published

Vol. 1: J. C. Erdmann. Wärmeleitung in Kristallen, theoretische Grundlagen und fortgeschrittene experimentelle Methoden. II, 283 Seiten. 1969.

Vol. 2: K. Hepp, Théorie de la renormalisation. III, 215 pages. 1969.

Vol. 3: A. Martin, Scattering Theory: Unitarity, Analyticity and Crossing. IV, 125 pages. 1969.

Vol. 4: G. Ludwig, Deutung des Begriffs „physikalische Theorie" und axiomatische Grundlegung der Hilbertraumstruktur der Quantenmechanik durch Hauptsätze des Messens. 1970. Vergriffen.

Vol. 5: Schaaf, The Reduction of the Product of Two Irreducible Unitary Representations of the Proper Orthochronous Quantummechanical Poincare Group. IV, 120 pages. 1970.

Vol. 6: Group Representations in Mathematics and Physics. Edited by V. Bargmann. V, 340 pages. 1970.

Vol. 7: R. Balescu, J. L. Lebowitz, I. Prigogine, P. Résibois, Z. W. Salsburg, Lectures in Statistical Physics. V, 181 pages. 1971.

Vol. 8: Proceedings of the Second International Conference on Numerical Methods in Fluid Dynamics. Edited by M. Holt. 1971. Out of print.

Vol. 9: D. W. Robinson, The Thermodynamic Pressure in Quantum Statistical Mechanics. V, 115 pages. 1971.

Vol. 10: J. M. Stewart, Non-Equilibrium-Relativistic Kinetic Theory. III, 113 pages. 1971.

Vol. 11: O. Steinmann, Pertubation Expansions in Axiomatic Field Theory. III, 126 pages. 1976.

Vol. 12: Statistical Models and Turbulence. Edited by C. Van Atta and M. Rosenblatt. Reprint of the First Edition. VIII, 492 pages. 1975.

Vol. 13: M. Ryan, Hamiltonian Cosmology. VII, 169 pages. 1972.

Vol. 14: Methods of Local and Global Differential Geometry in General Relativity. Edited by D. Farnsworth, J. Fink, J. Porter, and A. Thompson. V, 188 pages.

Vol. 15: M. Fierz, Vorlesungen zur Entwicklungsgeschichte der Mechanik. V, 97 Seiten. 1972.

Vol. 16: H.-O. Georgii, Phasenübergang 1. Art bei Gittergasmodellen. IX, 167 Seiten. 1972.

Vol. 17: Strong Interaction Physics. Edited by W. Rühl and A. Vancura. V, 405 pages. 1973.

Vol. 18: Proceedings of the Third International Conference on Numerical Methods in Fluid Mechanics, Vol. I. Edited by H. Cabannes and R. Temam. VII, 186 pages. 1973.

Vol. 19: Proceedings of the Third International Conference on Numerical Methods in Fluid Mechanics, Vol. II. Edited by H. Cabannes and R. Temam. VII, 275 pages. 1973.

Vol. 20: Statistical Mechanics and Mathematical Problems. Edited by A. Lenard. VIII, 247 pages. 1973.

Vol. 21: Optimization and Stability Problems in Continuum Mechanics. Edited by P. K. C. Wang. V, 94 pages. 1973.

Vol. 22: Proceedings of the Europhysics Study Conference on Intermediate Processes in Nuclear Reactions. Edited by N. Cindro, P. Kulišic and Th. Mayer-Kuckuk. XIV, 329 pages. 1973.

Vol. 23: Nuclear Structure Physics. Proceedings 1973. Edited by U. Smilansky, I. Talmi, and H. A. Weidenmüller. XII, 296 pages. 1973.

Vol. 24: R. F. Snipes, Statistical Mechanical Theory of the Electrolytic Transport of Non-electrolytes. V, 210 pages. 1973.

Vol. 25: Constructive Quantum Field Theory. The 1973 "Ettore Majorana" International School of Mathematical Physics. Edited by G. Velo and A. Wightman. III, 331 pages. 1973.

Vol. 26: A. Hubert, Theorie der Domänenwände in geordneten Medien. XII, 377 Seiten. 1974.

Vol. 27: R. K. Zeytounian, Notes sur les Ecoulements Rotationnels de Fluides Parfaits. XIII, 407 pages. 1974.

Vol. 28: Lectures in Statistical Physics. Edited by W. C. Schieve and J. S. Turner. V, 342 pages. 1974.

Vol. 29: Foundations of Quantum Mechanics and Ordered Linear Spaces. Advanced Study Institute, Marburg 1973. Edited by A. Hartkämper and H. Neumann. VI, 355 pages. 1974.

Vol. 30: Polarization Nuclear Physics. Proceedings 1973. Edited by D. Fick. IX, 292 pages. 1974.

Vol. 31: Transport Phenomena. Sitges International Schools of Statistical Mechanics, June 1974. Edited by G. Kirczenow and J. Marro. XIV, 517 pages. 1974.

Lecture Notes in Physics

Edited by J. Ehlers, München, K. Hepp, Zürich, and H. A. Weidenmüller, Heidelberg, and J. Zittartz, Köln
Managing Editor: W. Beiglböck, Heidelberg

51

Wolfgang Nörenberg
Hans A. Weidenmüller

Introduction to the Theory of Heavy-Ion Collisions

Springer-Verlag
Berlin Heidelberg GmbH 1976

Authors

Wolfgang Nörenberg
Hans Weidenmüller
Max-Planck-Institut für Kernphysik
Postfach 103980
6900 Heidelberg/BRD

Library of Congress Cataloging in Publication Data

Nörenberg, W   1938-
    Introduction to the theory of heavy-ion collisions.

    (Lecture notes in physics ; 51)
    Includes bibliographies and index.
    1. Heavy ion collisions.   I. Weidenmüller, Hans A.,
joint author. II. Title. III. Series.
QC794.6.C6N6   539.7'54   76-23166

ISBN 978-3-540-09753-2   ISBN 978-3-540-38271-3 (eBook)
DOI 10.1007/978-3-540-38271-3

Originally published by Springer-Verlag Berlin · Heidelberg · New York in 1976

This work would not have been possible without the help and encouragement of many people. Our ideas of how to present the material were clarified by critical questions of the participants in our lecture course. The rough draft of these notes was subject to detailed criticism by P. Armbruster, P. Braun-Munzinger, K. Dietrich, D. Fick, W. Frahn, R. Fuller, K. Gelbke, H.L. Harney, D. Pelte and W. von Oertzen, to all of whom we are very grateful. Needless to say, all remaining mistakes are ours. We are indebted to J. Knoll and R. Schaeffer for the permission to use unpublished material. Miss Browning, Mrs. Faulbaum and Mrs. Karl-Kratky typed various versions of the manuscript, and we owe thanks to each of them for their infinite patience and care.

These notes do not aim at a review of the field. We apologize to all authors whose work was not included, or cited incompletely or incorrectly. The manuscript was finished in September 1975, and no later developments were included.

Danksagung

As indicated in the captions, we have taken numerous figures (and some tables) from the literature. Thanks are due to the authors and the publishers for permission to do so.

Table of Contents

1.   Introduction.

With the advent of heavy-ion reactions, nuclear physics has acquired a new
frontier. The new heavy-ion sources operating at electrostatic accelerators and the
high-energy experiments performed at Berkeley, Dubna, Manchester and Orsay, have
opened up the field, and have shown us impressive new prospects. The new accelerators
now under construction at Berlin, Daresbury and Darmstadt, as well as those under
consideration (GANIL, Oak Ridge, etc.) are expected to add significantly to our
knowledge and understanding of nuclear properties. This applies not only to such
exotic topics as the existence and lifetimes of superheavy elements, or the possibility of shock waves in nuclei, but also to such more mundane issues as high-spin
states, new regions of deformed nuclei and friction forces. The field promises not
only to produce a rich variety of interesting phenomena, but also to have wide-spread
theoretical implications. Heavy-ion reactions are characterized by the large masses
of the fragments, as well as the high total energy and the large total angular
momentum typically involved in the collision. A purely quantum-mechanical description
of such a collision process may be too complicated to be either possible or interesting. We expect and, in some cases, know that the classical limit, the limit of
geometrical optics, a quantum-statistical or a hydrodynamical description correctly
account for typical features. We believe that a more precise understanding of the
limits of applicability of, and of the interplay between, these approaches will be
required by the data and will present a challenge to the theorists.

The present set of lecture notes, mainly the result of a course the authors
gave in the academic year 1973/4 at Heidelberg, is rather modest in scope, compared
with the hopes and expectations just mentioned. We have collected some basic tools
of theoretical heavy-ion physics which we believe to be important for future developments. In attempting to include primarily reasonably well-established results, we
have omitted a number of interesting topics which are rapidly developing, such as
the hydrodynamical approach and high-energy heavy-ion reactions.

We have aimed at a level of presentation which is intelligible to a graduate
student who is familiar with non-relativistic quantum-mechanics, and with the ele-

ments of nuclear physics. In trying to avoid lengthy derivations, we often give only the important physical arguments. In order to elucidate the physical content of the theoretical results, we have included a significant number of experimental data with which a comparison is made. We hope that the tables and graphs included in the text make these notes also useful in everyday applications.

Classical mechanics offers the simplest and, in some cases, as yet the only access to some results of heavy-ion collisions. This is described in chapter 2 . In chapter 3 , we collect a number of general features of heavy-ion collisions, relating to Q-values, grazing and critical angular momentum, Yrast levels and fusión, which determine many properties of cross sections. Chapter 4 presents some elements of nuclear scattering theory. In particular, the decomposition of the mean cross section into a shape elastic or direct part, and into a fluctuation or compound-nucleus part, is introduced, and the relationship between this decomposition and the contents of the following chapters is established. Chapter 5 is devoted to the elastic scattering of heavy-ions, one of the best-investigated topics of the field. Chapter 6 is a reminder of the theory of Coulomb excitation. Although excellent reviews of this field exist, we have included it here for two reasons. First, we wanted to emphasize features typical of heavy ions. Second, we found it convenient to base some of the theoretical developments in chapter 7, devoted to inelastic and transfer reactions, on the approximations introduced in the theory of Coulomb excitation. The nuclear-physics part of these notes concludes with chapter 8 in which we describe the statistical approach to the calculation of compound-nucleus reactions, precompound reactions, and deeply inelastic collisions.

Heavy-ion physics exceeds the frame of nuclear physics and relates also to atomic physics and solid-state physics, not to mention its applications in other fields of science and medicine. Because of the close similarity of many of the theoretical methods, and because of the strong link of the physical phenomena with those in nuclear physics, we have included in the (last) chapter 9 a survey of some of the problems in atomic physics encountered in heavy-ion collisions.

Throughout the text, we use the symbol HI as an abbreviation for heavy ions. References are collected alphabetically at the end of each chapter. Each chapter is divided into sections. Equations and figures are numbered consecutively in each section.

2. Classical theory of HI collisions

We consider the collision between two nuclei with mass numbers A_1 and A_2. The wave length of the relative motion at the top of the Coulomb barrier V_{CB} is given by

$$\lambdabar_{CB} = k_{CB}^{-1} \approx \sqrt{\frac{A_1 + A_2}{A_1 A_2} \frac{20}{E - V_{CB}}} \quad fm \tag{2.1}$$

where E denotes the incident center-of-mass energy and both E and V_{CB} are taken in MeV. Many HI collisions are investigated with heavy nuclei and/or high energies, such that the wave length is small as compared to characteristic lengths of the interaction potential. For example, $\lambdabar_{CB} \approx 0.1$ fm for Ar on Th at only 50 MeV above the Coulomb barrier. Therefore, these collisions can frequently be treated in the classical limit. In the first part of this section we develop the basic concepts and results of the classical theory of elastic scattering. This theory is considered in some detail because (i) the analogy with classical concepts is important later on, and (ii) the theory is easily generalized to apply to deeply inelastic collisions. The deeply inelastic collisions are treated in the second part of this chapter.

2.I. Elastic scattering

2.1 Classical deflection function and cross section

We consider the scattering of a particle (mass m) by a spherically symmetric real potential $V(r)$ in the non-relativistic limit. Introducing the spherical coordinates r, ϑ, φ to describe the trajectory of the particle, we get from the conservation of energy E and angular momentum L

$$\dot{r}^2 = \frac{2}{m} \left[E - V(r) - \frac{L^2}{2mr^2} \right] \ , \tag{2.1.1}$$

$$L = mr^2\dot{\vartheta} = const \ . \tag{2.1.2}$$

We eliminate dt from both equations and find

$$d\vartheta = -\frac{L}{mr^2} \left\{ \frac{2}{m} \left[E - V(r) - \frac{L^2}{2mr^2} \right] \right\}^{-\frac{1}{2}} dr \tag{2.1.3}$$

i.e., the differential equation of the left half of the trajectory which is illustrated in fig. 2.1.1 . Integration from r = ∞ to the minimal distance r_{min} gives the deflection angle

$$\Theta = \pi - 2 \int_{r_{min}}^{\infty} \frac{L}{mr^2} \sqrt{\frac{m/2}{E - V(r) - L^2/(2mr^2)}} \ dr \tag{2.1.4}$$

or after substituting w = b/r with the collision or impact parameter b = L(2mE)$^{-1/2}$

$$\Theta = \pi - 2 \int_{0}^{b/r_{min}} \frac{dw}{\sqrt{1 - V(w)/E - w^2}} \ . \tag{2.1.5}$$

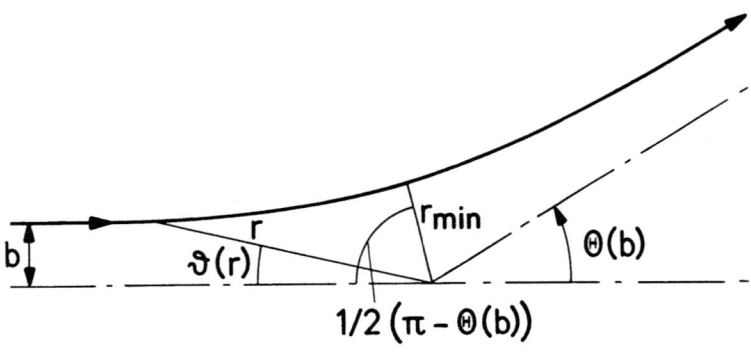

Fig. 2.1.1   Trajectory of a particle in a central field of finite range

Since the integrand is always positive, the deflection angle Θ can range from $-\infty$ to π. The scattering angle θ which ranges from 0 to π, is related to the deflection angle Θ by

$$\Theta + 2n\pi = \pm\theta \tag{2.1.6}$$

with integer n.

For a Coulomb potential $V(r) = Z_1 Z_2 e^2/r \equiv \frac{\alpha_c}{r}$ the evaluation of eq. (2.1.5) gives

$$\Theta = 2\arctan\frac{Z_1 Z_2 e^2}{2Eb} = 2\arctan\frac{\alpha_c}{2Eb} \tag{2.1.7}$$

which is illustrated in fig.2.1.2 . Here

$$r_{min} = \frac{\alpha_c}{2E}\left(1 + \sqrt{1 + (2Eb/\alpha_c)^2}\right) \tag{2.1.8}$$

is used which is obtained from energy and angular momentum conservation. Note that $r_{min} \geq \alpha_c/E$. If the point charge is replaced by an extended charge distribution

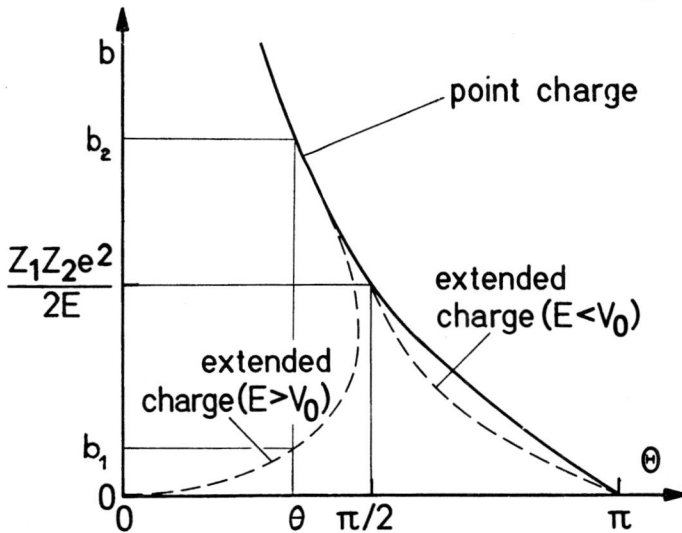

Fig.2.1.2 Deflection functions Θ(b) for the scattering of a charged particle from a point charge and an extended charge with total charge equal to the point charge and with a central potential height V_o .

then, in the head-on collision (b=0) and for energies E larger than the central potential V_o , the particle goes right through the center whereas the particle is reflected for b = 0 and $E < V_o$ to $\theta = \pi$. The corresponding deflection functions Θ (b) are also shown in fig. 2.1.2 . For partly or purely attractive potentials the deflection functions can become even more complicated.

The differential cross section $d\sigma/d\Omega$ is determined by the number of particles which are deflected per unit time into the solid angle $d\Omega$ divided by the incident flux N , i.e.

$$d\sigma = \frac{d\sigma}{d\Omega} \sin\theta \, d\theta \, d\phi = \frac{Nb \, |db| \, d\phi}{N} \qquad (2.1.9)$$

where we have to take the absolute value of the variation of b with positive $d\theta$. Hence, the classical cross section is

$$\left(\frac{d\sigma}{d\Omega}\right)_{c\ell} = \frac{b}{\sin\theta} \left|\frac{db}{d\theta}\right| = \frac{b}{\sin\theta} \left|\frac{db}{d\theta}\right| \qquad (2.1.10)$$

Inserting the Coulomb deflection function (2.1.7) we obtain the Rutherford cross section

$$\left(\frac{d\sigma}{d\Omega}\right)_{Ruth} = \frac{a_c^2}{4} \frac{1}{\left(\sin\frac{\theta}{2}\right)^4} \qquad (2.1.11a)$$

with

$$a_c \equiv \frac{\alpha_c}{2E} \qquad . \qquad (2.1.11b)$$

If two (b_1 and b_2 in fig. 2.1.2) or more impact parameters b_n lead to the same scattering angle θ , the classical cross section is given by summing the different contributions,

$$\left(\frac{d\sigma}{d\Omega}\right)_{c\ell} = \sum_n \frac{b_n}{\sin\theta} \left|\frac{db}{d\theta}\right|_{b=b_n} \qquad . \qquad (2.1.12)$$

Before we enter the discussion of classical scattering phenomena (rainbow, glory and spiral scattering) we first consider the heavy-ion interaction potential in some detail.

2.2 Heavy-ion interaction potentials

The interaction potential between two heavy ions consists of two parts, the Coulomb interaction which is rather well-known, and the nuclear part which is rather unknown. One may try to determine the nuclear interaction potential by a fit to experimental data, e.g. those for elastic scattering. But it turns out that, due to the interplay between nuclear attraction and absorption, the potential is not uniquely determined. Several methods have been introduced to calculate the real part of the nuclear interaction potential between colliding nuclei (for a review see [KR 75]) :

(i) In the folding procedure [MC 64, BR 74] one generates the interaction potential by folding together the single-particle density distribution ρ_1 of one nucleus with the single-particle potential V_2 of the other nucleus,

$$V_N^{12}(r) = \int d^3r_1 \, \rho_1(r_1) \, V_2(|\vec{r} - \vec{r}_1|) \ . \tag{2.2.1}$$

Since it turns out that $V_N^{12} \neq V_N^{21}$ one usually takes their mean value. A symmetric interaction potential is produced by folding a two-nucleon potential v into the projectile and target single-particle density distributions,

$$V_N^{12}(r) = V_N^{21}(r) = \int d^3x \int d^3y \, \rho_1(|\vec{x} - \vec{X}|) \, v(|\vec{x} - \vec{y}|) \, \rho_2(|\vec{y} - \vec{Y}|) \tag{2.2.2}$$

where $\vec{r} = \vec{X} - \vec{Y}$ is the vector connecting the centers $\vec{X}$ and $\vec{Y}$ of the nuclei.

(ii) The energy density $\varepsilon(\rho)$ has been used [BR 68, SC 69, LO 73, NG 75] to obtain the nuclear interaction potential from the integral

$$V_N^{12}(r) = \int d^3r' \left[\varepsilon(\rho_1 + \rho_2) - \varepsilon(\rho_1) - \varepsilon(\rho_2) \right] \ . \tag{2.2.3}$$

Here $\varepsilon(\rho)$ is the nuclear binding energy per unit volume calculated at the density ρ. Generally, $\varepsilon(\rho)$ depends not only on the density ρ but also on derivatives of ρ (surface effects).

(iii) A useful approximation has been introduced by the proximity potential [RA 74]. Performing the integrals in eqs. (2.2.2) and (2.2.3) up to an integration dxdy over the plane $\mathscr{S}$ perpendicular to the vector which connects the fragment centers we have

$$V_N^{12}(r) = \int_{\mathscr{S}} dx\,dy \; e(x,y;r) \tag{2.2.4}$$

where $e(x,y;r)$ denotes the interaction energy per unit area at the point x,y on the surface. This quantity $e(x,y;r)$ generally depends on the range γ of the nuclear interaction and on the radii R_i and diffusenesses a_i of the nuclear density distributions. For $R_i \gg a_i$ (leptodermous systems) and $R_i \gg \gamma$ the dependence on R_i is negligible and hence, $e(x,y;r)$ becomes approximately equal to the interaction energy per unit area between two flat nuclear surfaces. Thus, $e(x,y;r) = e_\infty(\xi'(x,y))$ where ξ' is the distance between the nuclear surfaces. Introducing ξ' as a variable and integrating over the other variable leads to

$$V_N^{12}(r) \approx V_{prox}(\xi) = 2\pi \frac{R_1 R_2}{R_1 + R_2} \int_\xi^\infty e_\infty(\xi')\,d\xi' \tag{2.2.5}$$

where R_1, R_2 denote the radii and $\xi = r - R_1 - R_2$ the distance between the surfaces of the interacting nuclei. The universal function $e_\infty(\xi')$ can be calculated by methods (i) or (ii).

(iv) Self-consistent methods (Hartree-Fock or Hartree-Fock-Bogoljubov calculations) can also be applied to calculate the interaction potentials between colliding nuclei.

In the methods (i) to (iii) the densities of the nuclei are assumed to be fixed during the collision ("frozen density approximation"). Hence, it is assumed that the nucleons do not readjust during the interaction time (therefore also "sudden approximation"). It is not clear to what extent this assumption is reasonable for those HI collisions where the kinetic energy per particle is only 1 to 2 MeV at the Coulomb barrier. Method (iv) implies the opposite assumption, i.e., the nucleons adjust their motion such that for any distance r between the centers, the total energy of the system is a minimum ("adiabatic approximation"). In general, the HI interaction potential must be velocity dependent. Another problem which arises is connected with the number of degrees of freedom that have to be taken into account explicitly.

Aside from calculating HI potentials according to some prescription one may also choose a reasonable form, e.g. a Woods-Saxon potential

$$V_N(r) = - V_o \left(1 + exp \frac{r-R}{a} \right)^{-1} \tag{2.2.6}$$

with adjusted parameters V_o, R and a .

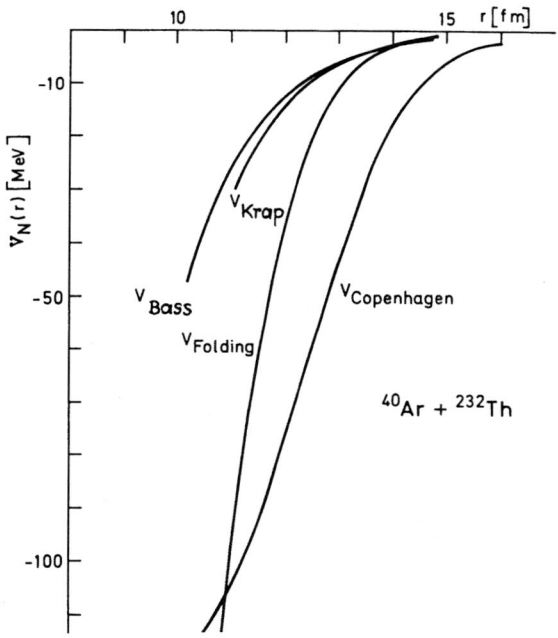

Fig. 2.2.1 Various nuclear interaction potentials used for ^{40}Ar + ^{232}Th. V_{Bass}, V_{Krap} and $V_{folding}$ are obtained with different folding procedures. $V_{Copenhagen}$ is a Woods-Saxon potential with adjusted parameters. (From [GR 75b]).

Although we must have severe doubts about the nuclear part of the HI potentials, it was extremely useful to use such potentials in connection with the
Coulomb potential to study at least qualitatively the elastic scattering and deeply
inelastic collisions (cf. sections 2.5 to 2.9) between heavy ions. Fig. 2.2.1 illustrates the variety of nuclear potentials which have been used for example in the
^{40}Ar + ^{232}Th collision. For fixed angular momentum L = $\hbar\ell$ the total effective HI
potential $V_\ell(r)$ is the sum of the nuclear part $V_N(r)$, the Coulomb part $V_{Clb}(r)$
and the centrifugal barrier $L^2/(2mr^2)$ with m denoting the reduced mass of the
nuclei. Fig. 2.2.2 gives an example of a typical HI potential in the frozen density
approximation where a repulsive core evolves from the high particle densities
(compression) in the overlap region. For small ℓ values ($\ell = L/\hbar$) these potentials
have pockets. At a critical ℓ value (ℓ_{crit}) between ℓ = 80 and 120 these pockets
vanish. The resulting forces being partially repulsive and partly attractive,
deflection functions as illustrated in fig. 2.2.3 are obtained from eq. (2.1.5).

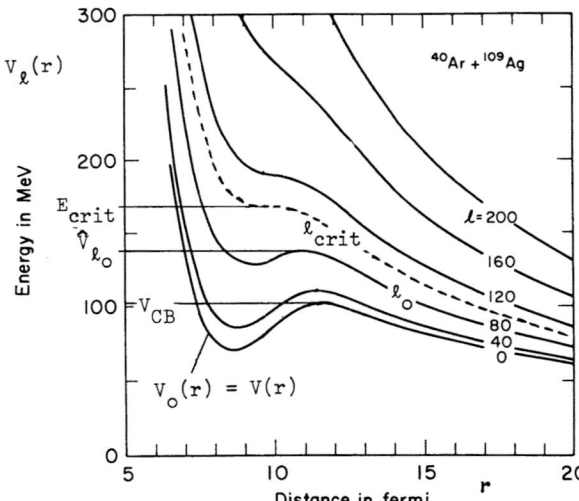

Fig. 2.2.2 Effective nucleus-nucleus potential $V_\ell(r)$ for ^{40}Ar + ^{109}Ag
in the frozen density approximation. The nuclear part of the
interaction potential has been obtained from the expression
(2.2.5) with $e_\infty(\xi')$ calculated from a Thomas-Fermi approximation using an effective nucleon-nucleon interaction [RA 74].
(From [MY 74]). For further details see subsection 2.3.3 .

For large impact parameters b the deflection functions coincide with the Coulomb
deflection function (2.1.7) shown in fig. 2.1.2 . With decreasing b the particle
gets into the range of the attractive nuclear potential. The trajectory is bent
forward towards negative deflection angles. For a given potential it depends on
the incident energy how negative the deflection angle can become. For even smaller
b the particle begins to feel the repulsive core of the potential and the deflection
angle increases with decreasing b (as for Coulomb scattering). We shall discuss
in the following subsection the scattering phenomena (rainbow, glory, orbiting)
which are connected with the type of deflection functions shown in fig. 2.2.3 .

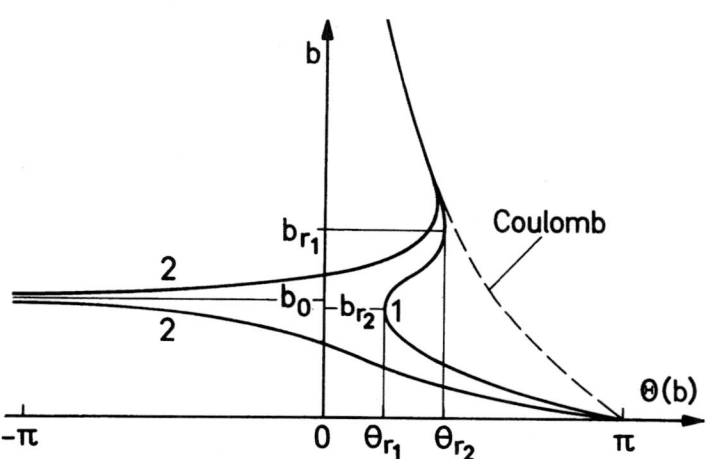

Fig. 2.2.3 Typical deflection functions obtained for partly
repulsive and partly attractive forces (cf. fig.2.2.2).

2.3   Rainbow, glory and spiral scattering

We shall discuss these classical phenomena somewhat in detail because they are important also in deeply inelastic collisions. Rainbow, glory and spiral scattering are due to the presence of more than one trajectory leading to the same scattering angle.

2.3.1 Rainbow scattering

This type of scattering occurs whenever $d\theta/db = 0$. The angle θ_r where the derivative vanishes is called the rainbow angle. In the deflection function 1 of fig. 2.2.3 , there are two rainbows. The vanishing of the derivative has two consequences: (i) The stationarity of θ with b implies that many particles with slightly different impact parameters are focussed into the same scattering angle. The cross section is thus expected to be very large. Eq. (2.1.12) shows, indeed, that at $\theta = \theta_r$, the cross section becomes infinite. (ii) In the vicinity of θ_r, the available scattering angles are limited from above (θ_{r_2} in fig. 2.2.3) or below (θ_{r_1} in fig. 2.2.3), and it makes sense to speak of the "bright side" and the "dark side" of a rainbow.

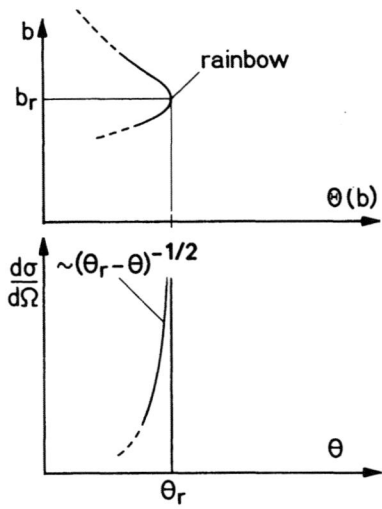

Fig. 2.3.1 Contribution to the total cross-section from a rainbow.

Around the rainbow angle we have the expansion

$$\theta = \theta_r + q \, (b - b_r)^2 \tag{2.3.1}$$

and hence $d\theta/db = 2q(b - b_r)$. Inserting this into (2.1.12) and using eq. (2.3.1) again, we obtain the classical cross-section

$$\left(\frac{d\sigma}{d\Omega}\right)_r = \frac{b_r}{\sin\theta} \, \frac{1}{\sqrt{q\,(\theta - \theta_r)}} \tag{2.3.2}$$

near the rainbow angle and on the bright side of it. Note that the two branches of the deflection function for $b > b_r$ and $b < b_r$ give almost equal contributions to the cross section. The contribution from a rainbow to the total cross section is illustrated in fig. 2.3.1 . Contributions from other branches of the deflection function have to be added to $(d\sigma/d\Omega)_r$ to obtain the total classical cross section.

2.3.2 Glory scattering

This type of scattering occurs whenever the deflection angle $\theta(b)$ takes the value $-n\pi$ (with n integer) for $b_n \neq 0$. According to eq. (2.1.12), the classical cross section becomes infinite because $\sin\theta$ vanishes at these points. There is a backward glory for $n = 1,3,5, \ldots$ and a forward glory for $n = 0,2,4, \ldots$. Since for spherically symmetric potentials the deflection angle is always smaller than π for $b \neq 0$, there is no glory scattering for $n = -1$. As an example we consider a forward glory around $\Theta = \theta = 0$. In the deflection function 2 of fig. 2.2.3 two points with $\Theta_n = 0$ for $b_n \neq 0$ occur. All particles incident with impact parameters $b_n - db < b < b_n + db$ are scattered into forward directions $0 < \theta < d\theta$ thus giving rise to the geometrical singularity $1/\sin\theta$. Close to 0 the deflection function can be approximated by

$$\Theta(b) = \zeta \, (b - b_{gl}) \tag{2.3.3}$$

yielding the sum of two almost equal contributions for $\Theta > 0$ and $\Theta < 0$,

$$\left(\frac{d\sigma}{d\Omega}\right)_{gl} = \frac{2b_{gl}}{|\zeta|} \frac{1}{\sin\theta} \tag{2.3.4}$$

for the classical cross section near $\theta = 0$.

2.3.3 Spiral scattering (orbiting)

Suppose that $V(r) = V_N(r) + V_{Clb}(r)$ has a maximum V_{CB} at $r = R$ as shown in fig. 2.2.2 . The effective potentials $\hat{V}_\ell(r)$ exhibit corresponding maxima V_ℓ up to a critical angular momentum ℓ_{crit}. For energies $E < V_{CB}$ there is a classical turning point for all ℓ (or b). Spiral scattering or orbiting occurs for scattering energies $V_{CB} < E < \hat{V}_{\ell_{crit}}$. For a fixed energy E , the deflection function as function of ℓ (or b) becomes singular around ℓ_o (or b_o) which is determined by $\hat{V}_{\ell_o} = E$. The particle incident with ℓ_o(or b_o) approaches the maximum $\hat{V}_{\ell_o}$ in an infinitely long time, rotating (or orbiting) with constant angular momentum ℓ_o around the origin. The resulting trajectory is a spiral (hence, spiral scattering). In order to derive the singular behaviour of Θ (b) near $b = b_o$, one has to consider the integral of eq. (2.1.5) in some detail. The result is that Θ (b) exhibits a logarithmic singularity at $b = b_o$, as is indicated in fig. 2.2.3 for the deflection function 2 . From the vicinity of b_o a rather isotropic cross section results. For energies $E > \hat{V}_{\ell_{crit}}$ the singularity in the deflection function changes to a rainbow as shown by the deflection function 1 in fig. 2.2.3 .

2.3.4 Application to elastic heavy-ion scattering

It is obvious from fig. 2.1.2 that only rainbow scattering can occur in a purely repulsive potential. All three kinds of classical scattering phenomena can occur in a purely attractive potential as well as in a combined (repulsive and attractive) potential. The following example is taken from Broglia et al. [BR 72]. These authors study the scattering of ^{16}O on ^{58}Ni at E_{lab} = 60 MeV by assuming three different real Woods-Saxon potentials in addition to the Coulomb potential. The parameters (cf. eq.(2.2.6) with $R = r_o(A_1^{1/3} + A_2^{1/3})$) are as follows: (I)$V_o$ = 2.0 MeV ,

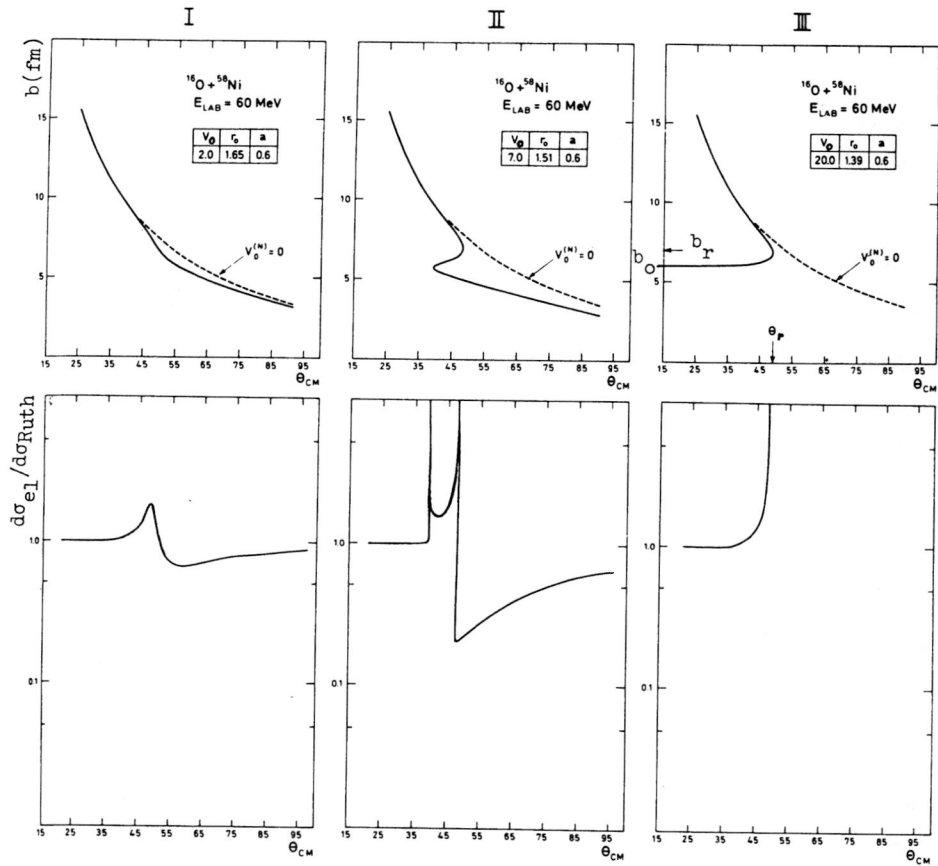

Fig. 2.3.2 Deflection function and angular distribution for
^{16}O(60 MeV) + ^{58}Ni. (From [BR 72]).

a = 0.6 fm, r_o = 1.65 fm; (II) V_o = 7.0 MeV, a = 0.6 fm, r_o = 1.51 fm; (III) V_o =
20.0 MeV, a = 0.6 fm, r_o = 1.39 fm. All three potentials have the same asymptotic
behaviour for large r given by the folded potential obtained from the symmetrized
expression (2.2.1). The deflection functions and the corresponding angular distributions are plotted in fig. 2.3.2 . Potential I corresponds to a weak nuclear attraction where the trajectory is only changed slightly from the Coulomb trajectory.
The maximum in the angular distribution is due to a focussing effect of the real
potential as shown in the deflection function. Potential II corresponds to an attraction of intermediate strength and gives rise to two rainbow singularities.
Potential III corresponds to a strong nuclear attraction which gives rise to a
maximum of the potential. Therefore, both rainbow and spiral scattering is expected
in this case. But only the rainbow part has been calculated because for b < b_o the
scattering would depend strongly on the potential in the nuclear interior. This
part of the potential is strongly absorptive and not well known (cf. section 5.2).

2.4   Limitations due to quantal effects and absorption

There are two effects which limit the classical description of elastic
scattering. These are quantal effects and absorption. As we shall see in section 5 ,
both the quantal effects and the absorption do not destroy the classical picture
completely. At large energies the deflection function remains a useful tool for
describing the elastic scattering, and the scattering phenomena, like rainbow or
glory scattering, are still there although they are significantly modified due to
interference effects.

2.4.1 Limitations due to quantal effects

The transition from the Schrödinger equation to the Hamilton-Jacobi equation
of classical mechanics can be performed as follows. Let

$$i\hbar \frac{\partial}{\partial t} \Psi(\vec{r},t) = \left\{ -\frac{\hbar^2}{2m} \Delta + V(\vec{r}) \right\} \Psi(\vec{r},t) \qquad (2.4.1)$$

be the Schrödinger equation for a single-particle in a local potential $V(\vec{r})$.

We put

$$\Psi(\vec{r},t) = B(\vec{r},t) \exp\left[iS(\vec{r},t)/\hbar\right] \tag{2.4.2}$$

where B and S are real. Inserting this into eq. (2.4.1) and taking the real and imaginary parts of the resulting equation, we find

$$\frac{\partial S}{\partial t} + \frac{1}{2m}(\vec{\nabla}S)^2 + V(\vec{r}) = \frac{\hbar^2}{2m}\frac{\Delta B}{B} \quad , \tag{2.4.3}$$

$$m\frac{\partial}{\partial t}B + (\vec{\nabla}B)\cdot(\vec{\nabla}S) + \tfrac{1}{2}B\,\Delta S = 0 \; . \tag{2.4.4}$$

It is easily verified that eq. (2.4.4) is the continuity equation $\dot{\varrho} + div\,\vec{j} = 0$ appropriate to the form (2.4.2) of $\Psi(r,t)$. Hence, eq. (2.4.3) contains the dy-- namics of the problem. Formally, the classical equation is obtained by putting $\hbar = 0$ in eq. (2.4.3). This yields

$$\frac{\partial}{\partial t}S_{cl} + \frac{1}{2m}\left(\vec{\nabla}S_{cl}(\vec{r},t)\right)^2 + V(\vec{r}) = 0 \; , \tag{2.4.5}$$

the Hamilton-Jacobi equation for the classical action function $S_{cl}(\vec{r},t)$, often called Hamilton's principal function. Since the potential $V(r)$ is independent of time, we have

$$S_{cl}(\vec{r},t) = W(\vec{r}) - Et \; . \tag{2.4.6}$$

The interpretation of $W(r)$ and E is obtained by observing that S_{cl} is the in- definite integral of the Lagrangian,

$$S_{cl} = \int dt \, \mathcal{L} \tag{2.4.7}$$

along a classical trajectory $r(t)$. Identifying E with the sum of the kinetic energy T and the potential energy V, we have $\mathcal{L} = T - V = 2T - E$, so that

$$W(\vec{r}) = \int_0^t dt' \, m\left[\dot{\vec{r}}(t')\right]^2 = \sum_{j=1}^{3} \int_{x_j=0}^{x_j(t)} dx_j \, p_j \quad , \tag{2.4.8}$$

which implies grad $W = \vec{p}$ where $\vec{p}(\vec{r})$ is the local momentum. We introduce the local wave length

$$\lambdabar(\vec{r}) = \frac{\hbar}{\sqrt{2m\,[E - V(\vec{r})]}} = \frac{\hbar}{|\vec{p}(\vec{r})|} \quad . \tag{2.4.9}$$

Then, it follows from eqs. (2.4.3) and (2.4.6) that, for stationary problems where (2.4.6) is valid also quantum-mechanically,

$$(\vec{\nabla} S)^2 = \frac{\hbar^2}{\lambdabar^2}\left(1 + \lambdabar^2\,\frac{\Delta B}{B}\right). \tag{2.4.10}$$

The omission of $\lambdabar^2\Delta B/B$ in this equation gives the classical result and is justified if

$$\lambdabar^2(\vec{r})\left|\frac{\Delta B}{B}\right| \ll 1 \quad . \tag{2.4.11}$$

More precisely, the regions where eq. (2.4.11) is violated must be negligibly small, so that the solution of $(\vec{\nabla} S)^2 = \hbar^2/\lambdabar^2$ gives a sufficiently good approximation to the solution of eq. (2.4.10). The condition (2.4.11) is always fulfilled for $V = 0$ and plane waves, for which $B = $ constant. In contradistinction, (2.4.11) may be violated, for $V = 0$, if we choose a wave packet concentrated within a narrow region, because of the quantum-mechanical spreading of the wave packet. Condition (2.4.11) is always violated if interference phenomena are present, because the sum of two wave functions (2.4.2) can be expressed in the same form with B and S real, only if $B(r,t)$ is a strongly oscillating function. The condition (2.4.11) imposes also a restriction on the r-dependence of the potential,

$$\left|\text{grad }\lambdabar(\vec{r})\right|^2 \ll 1 \quad , \tag{2.4.12}$$

which is obtained from condition (2.4.11) in the following way. In the stationary case, eq. (2.4.4) yields a qualitative information on $\vec{\nabla} B$, obtained by replacing S by S_{cl} and by assuming that $\vec{\nabla} B$ is in the direction of $\vec{\nabla} S_{cl} = \vec{p}$ (This last assumption is consistent with the condition that the transverse dimensions of the wave, and characteristic length of the transverse variation of B, are large compared to λ). It then follows from eq. (2.4.4) that

$$\frac{\Delta B}{B} = -\frac{1}{2}\, div\left(\frac{\vec{p}}{p^2}\,div\,\vec{p}\right) + \frac{1}{4}\left(\frac{1}{p}\,div\,\vec{p}\right)^2 \approx \frac{3}{4}\,p^2\left(grad\,\frac{1}{p}\right)^2. \qquad (2.4.13)$$

The last form is obtained if we assume that the transverse change of $\vec{p}$ can be neglected, and that second derivatives of $\vec{p}$ are negligible. The result, combined with eq. (2.4.11) gives the condition (2.4.12). We apply this condition to the scattering by a heavy-ion potential. Let M be the nucleon mass, m the reduced mass, and define A_{red} by $m = A_{red}M$. Then, from eq. (2.4.9) we find

$$\left|grad\,\lambda\right|^2 = \frac{\hbar^2}{8A_{red}M}\frac{(grad\,V)^2}{[E-V(\vec{r})]^3} \approx \frac{\hbar^2}{8A_{red}M}\frac{(V_0/d)^2}{(E-V_{CB})^3} \qquad (2.4.14)$$

where we assume the Coulomb potential to be smoothly varying, such that only grad $V_N(r)$ has to be considered. We have approximated the gradient of the nuclear potential by the ratio of the depth V_0 over the surface thickness d, $(E-V(r))$ by $E - V_{CB}$, where V_{CB} is the height of the Coulomb barrier. The latter approximation is valid if E is larger than V_{CB}. Using $V_0/d = 100$ MeV fm^{-1} we find from (2.4.12) the condition

$$A_{red}\left(E - V_{CB}\right)^3 \gg 5 \cdot 10^4 \ MeV^3. \qquad (2.4.15)$$

This condition is easily fulfilled for energies sufficiently far above the Coulomb barrier.

For a pure Coulomb potential $V(r) = \alpha/r$ we obtain

$$\left|grad\,\lambda\right|^2 = \left|\frac{d}{dr}\,\lambda\right|^2 = \frac{1}{\eta^2}\frac{a_c^4}{r\,(r-2a_c)^3} \qquad (2.4.16)$$

where $a_c \equiv \alpha/(2E)$ and the Sommerfeld parameter

$$\eta = \frac{Z_1 Z_2 e^2}{\hbar v} = \frac{\alpha_c}{\hbar v} \overset{(2.1.7)}{=} \frac{L}{\hbar}\,tan\,\frac{\theta}{2} \qquad (2.4.17)$$

is used. The quantities $v = (2E/m)^{1/2}$ and $L = mvb$ are the asymptotic velocity and the angular momentum of the particle. Quantitatively, a convenient formula for η is

$$\eta = 0.16 \, Z_1 Z_2 \sqrt{\frac{A_{projectile}}{E_{lab}}} = 0.16 \, Z_1 Z_2 \sqrt{\frac{A_{red}}{E_{(CM)}}} \qquad (2.4.18)$$

where E_{lab} and $E_{(CM)}$ are in MeV.

The condition (2.4.12) together with eqs. (2.1.7), (2.1.8) and (2.4.16) gives

$$\eta^2 \gg \eta^2_{crit} \quad where \quad \eta_{crit} = \frac{\sin^2 \theta/2}{(\cos \theta/2)(1 - \sin \theta/2)} \qquad . \qquad (2.4.19)$$

The dependence of η_{crit} is shown in Table 2.4.1 . Obviously this condition can never be met at backward angles, while it essentially implies $\eta \gg 1$ at smaller angles. One might imagine that the classical approximation breaks down only when the two projectiles approach each other so closely that nuclear forces come into play. This is not the case, however, since the distance of closest approach r_{min} may well exceed the sum of the radii of the two ions and yet $\eta < \eta_{crit}$.

Table 2.4.1

Values for η_{crit} as function of the scattering angle θ

θ	$0°$	$10°$	$20°$	$30°$	$40°$	$50°$	$60°$	$70°$	$80°$
η_{crit}	0	0.008	0.04	0.09	0.19	0.34	0.57	0.94	1.50

$90°$	$100°$	$110°$	$120°$	$130°$	$140°$	$150°$	$160°$	$170°$	$180°$
2.41	3.90	6.47	11.2	20.7	42.8	105	368	3000	∞

The modifications due to quantal effects are treated in the semiclassical approximation in section 5.1 . As is shown there, the quantum-mechanical scattering phase shifts can be simply related to the classical deflection function when the conditions for a semi-classical description of the quantum-mechanical scattering are met. The main difference to the classical treatment is due to interference effects from different trajectories. Instead of summing up the cross sections as done in eq. (2.1.12) one has to sum the amplitudes for the contributing trajectories. Therefore, the classical rainbow and glory scattering are modified significantly, but the behaviour of the cross section can still be interpreted with the help of the classical deflection function.

2.4.2 Limitations due to absorption

We consider a classical trajectory passing through the interaction region. Along the path both the projectile and the target may be excited. This gives rise to a reduction of the elastic cross section and is referred to as absorption. In order that the classical path remains unchanged and that only the reduction of probability along the trajectory has to be taken into account, the absorption has to be small over the spatial extension of a wave length. Strong absorption would imply strong reflection. The latter has never been observed in HI reactions. It, therefore, seems to be justified to study the case of weak absorption. Since usually $\lambdabar$ << interaction radius R in HI scattering, the absorption along the total classical path can still become very large. For light nuclei like $^{16}O + ^{16}O$ the absorption is rather weak up to the point of orbiting because only for $b < b_o$ do the trajectories lead over the Coulomb barrier and thus penetrate into the strongly absorptive region. This situation is illustrated in fig. 2.4.1(a). Around the Coulomb rainbow, the absorption is small. Apart from quantal modifications (cf. section (5.1), the angular distribution of the elastic cross section is well described by the refraction of trajectories giving rise to rainbow scattering around θ_r . For heavier nuclei the situation changes qualitatively: The Coulomb rainbow is due to the focussing of the Coulomb trajectories into forward directions by the attractive nuclear interaction. Since, for heavier nuclei, the Coulomb repulsion becomes larger, the nuclear attraction has to be

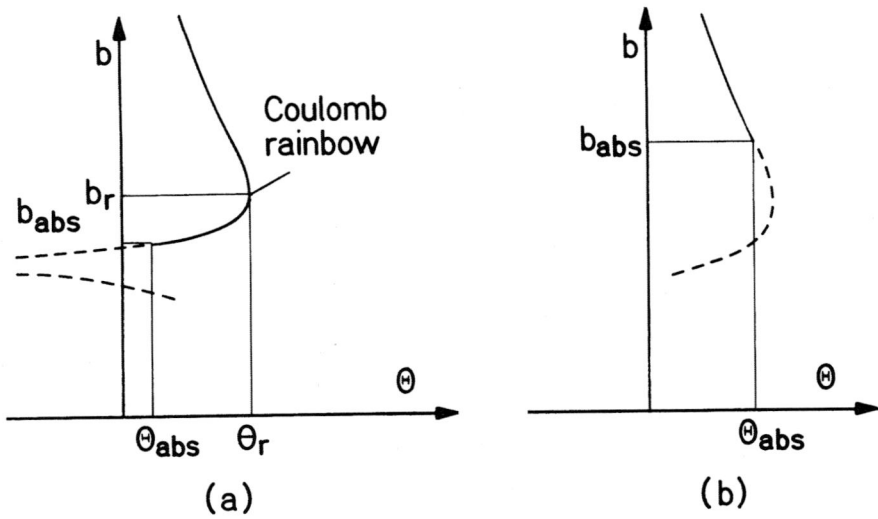

Fig. 2.4.1 Illustration of the onset (Θ_{abs}, b_{abs}) of strong absorption
for the collision of (a) light nuclei like $^{16}O + ^{16}O$ and
(b) heavy nuclei like $^{238}U + ^{238}U$.

stronger, too, in order to produce a Coulomb rainbow. Therefore, apart from stronger

Coulomb excitation, the absorption due to transitions induced by the nuclear inter-

action is expected to be much stronger, too. This situation is illustrated in

fig. 2.4.1(b) . The corresponding modification of the classical cross section is

treated within the diffraction model in sections 5.3 and 5.4 .

2.II. Deeply inelastic collisions

2.5   Experimental situation

Several attempts have been made in the past few years at Berkeley, Dubna and

Orsay to produce superheavy elements ($Z \gtrsim 114$) by bombarding suitable targets with

heavy ions like Ar, Kr, Xe and Ge . No superheavy element has been detected until

now. Nevertheless, these experiments have led to the discovery of a new and inter-

esting kind of reaction which is frequently referred to as deeply inelastic collision

(DIC), quasi-fission, incomplete fusion or strongly damped collision [AR 73, LE 73,

HA 74, MO 74, VO 74, WO 74, AL 75, CO 75, JA 75, PE 75, VO 75] .

For a recent review see $\left[\text{FL 74}\right]$. Some typical results of the experiments ^{40}Ar(297, 388 MeV) + ^{232}Th $\left[\text{AR 73}\right]$, ^{63}Cu(365 MeV) + ^{197}Au $\left[\text{PE 75}\right]$ and ^{84}Kr(600 MeV) + ^{209}Bi $\left[\text{WO 74}\right]$ are shown in figs. 2.5.1 to 2.5.6 .

Fig. 2.5.1 shows some typical results obtained in the collision ^{40}Ar(388 MeV) + ^{232}Th . Reaction products with masses close to Ar are seen to be produced within a broad range of energies. Generally, the spectra consist of two contributions, a high energy part and a low energy part. The high energy part has been attributed $\left[\text{AR 73, WI 73}\right]$ to a positive deflection angle $\Theta = \theta$, and the low energy part to the corresponding negative deflection angle , $\Theta = -\theta$. It is assumed that in order to reach the negative deflection angles the nuclei stay together for a longer time than would be needed for scattering into positive deflection angles. This is con- sistent with the change of the high energy peak towards smaller energies as one goes from the grazing angle to more forward angles. Elements between oxygen and vanadium have been observed. For products which are far from Ar, only the low energy part is seen. The lower end of the spectra is close to the energy of the Coulomb barrier for the fragments. The energy turns out to be somewhat smaller than that calculated for two touching spherical fragments. This has been attributed to a process in which

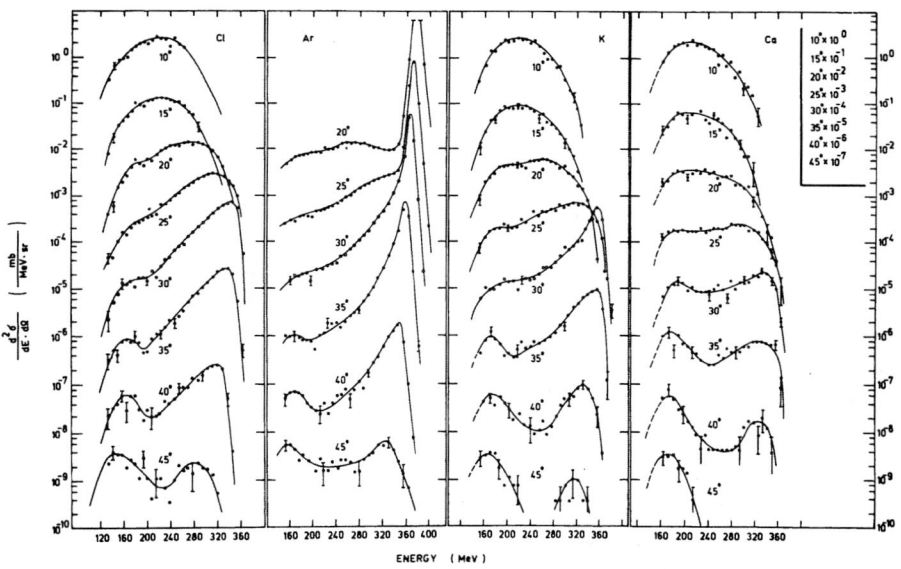

Fig. 2.5.1   Energy spectra for different lab. angles of Cl, Ar, K and Ca produced in the bombardment of ^{232}Th by 388 MeV ^{40}Ar . (From $\left[\text{AR 73}\right]$).

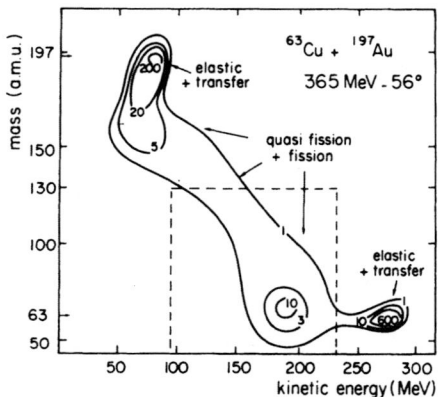

Fig. 2.5.2 Contour plot of the mass and energy distribution of the
reaction products in ^{63}Cu + ^{197}Au. (From [PE 75]).

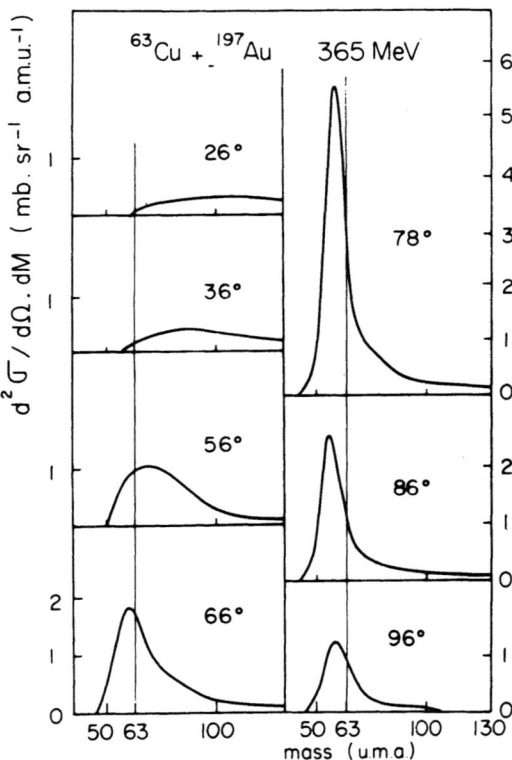

Fig. 2.5.3 Mass distributions of the light quasi-fission and fission
products at different scattering angles. The corresponding
kinetic energies are close to the Coulomb barrier. (From [PE 75]).

the two HI partly coalesce during the collision. The combined system deforms in a fission-like way as the two fragments separate.

Figs. 2.5.2 to 2.5.4 show results for the collision ^{63}Cu(365 MeV) + ^{197}Au . For a fixed angle of observation the observed mass and energy distributions are shown in Fig. 2.5.2 . The quasi-elastic peaks (elastic + transfer) for the two complementary masses are clearly separated from the low energy part which is denoted by 'quasifission + fission' in the diagram. The 'quasifission + fission' part is displayed in fig. 2.5.3 as mass distributions (by integration over the energies in the region denoted by the dashed square in fig. 2.5.2) for different angles of observation. Close to the grazing angle, $\theta_{lab} = 78^{\circ}$, there is a very sharp mass distribution which is peaked about 4 mass units below the projectile mass. This may be attributed to the decay of the excited fragment by particle emission. Towards forward angles the mass distributions becomes broader, indicating a larger interaction time. Even symmetric mass fragmentation becomes significant which is interpreted as compound-nucleus fission. Assuming a mass distribution for the fission component which is consistent with the excitation energy, the total distribution has been divided into a fission and a quasi-fission (deeply inelastic collision) part, cf.fig.2.5.4 .

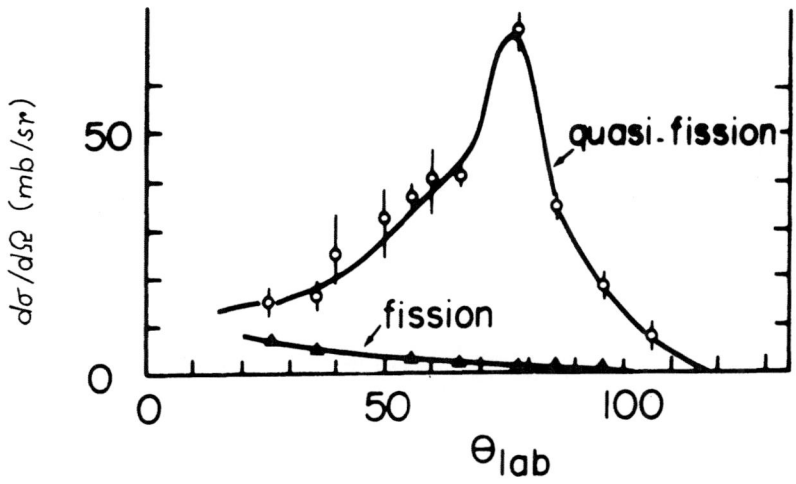

Fig. 2.5.4 Angular distribution of quasi-fission and fission events. (From [PE 75]).

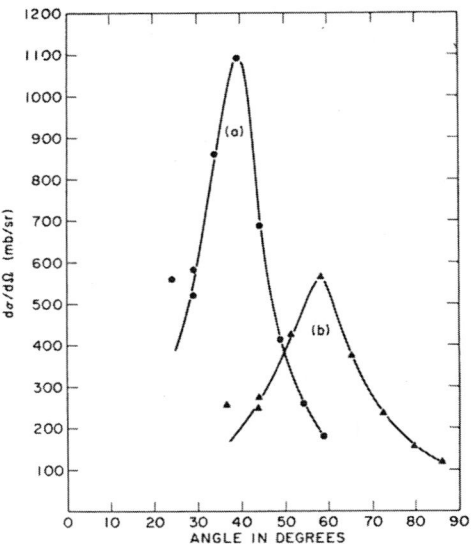

Fig. 2.5.5 Angular distribution of the light-mass fragment from the
strongly damped collisions in the bombardment of ^{209}Bi
with 600 MeV ^{84}Kr. The curves (a) and (b) correspond to
the lab and c.m. system, respectively. (From [WO 74]).

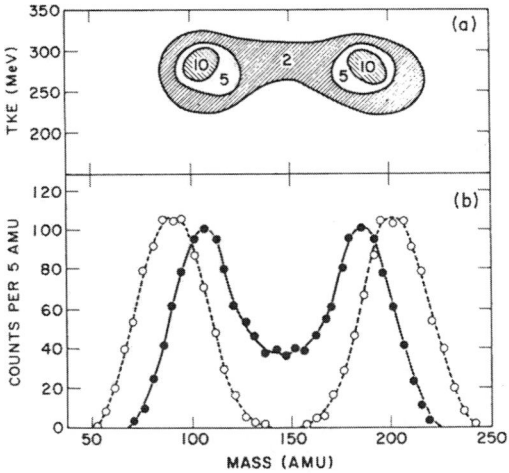

Fig. 2.5.6 (a) Contour plot of the kinetic energy (c.m.) versus fragment
mass at 59° (lab) for the strongly damped collisions.(b) Mass distributions for strongly damped collisions at θ_{lab} = 59° (solid curve)
and θ_{lab} = 34° (dashed curve). (From [WO 74]) .

Figs. 2.5.5 and 2.5.6 show similar results for the reaction ^{84}Kr (600 MeV)+ ^{209}Bi . The angular distribution for the strongly damped collisions (cf. fig.2.5.5) show a similar peak as that for ^{63}Cu + ^{197}Au . The mass and energy distribution (cf. fig.2.5.6) show that the fragment energies roughly correspond to the Coulomb energy and that a considerable amount of mass is transferred in these strongly damped collisions.

The experimental results are summarized as follows:

(i)   Projectile and target are mostly rather heavy nuclei with mass numbers $A \gtrsim 40$.

(ii)   The incident energy is large, typically 1 to 2 MeV per nucleon above the Coulomb barrier.

(iii) For increasing masses of the colliding nuclei and/or increasing energy the cross section for deeply inelastic collisions increases.

(iv) The angular distribution is strongly non-isotropic with the characteristic properties of a peripheral fast collision, the interaction time τ_{int} being of the order of 10^{-21}sec (i.e. $\leq$ rotational time of the composite system).

(v)   Despite this short interaction time a large fraction of the relative kinetic energy is dissipated into internal excitations of the colliding nuclei and events with a large number of transferred nucleons are observed. Recently, it has been observed [AL 75] that also a considerable amount of relative angular momentum is dissipated into internal angular momentum of the fragments.

(vi) Since the number of open channels is extremely large at these high excitation energies, it is impossible to resolve the individual channels and hence, only averaged quantities are observed.

2.6 Phenomenological description. Friction

The main efforts in describing and understanding the deeply inelastic collisions have been concentrated on the study of the relative motion by introducing dissipative terms (friction forces) into the classical equations of motion. The introduction and justification of friction forces together with some numerical results for deeply inelastic collisions are described in the following parts of this chapter. We expect that the understanding of this process is the key for understanding the total reaction mechanism in the collisions between really heavy nuclei.

Friction forces have been introduced $\begin{bmatrix} BO\ 74,\ GR\ 74,\ TS\ 74 \end{bmatrix}$ in order to describe the dissipation of relative kinetic energy and relative angular momentum within the classical equations of motion. In such a description no attention is paid to the transfer of mass and charge. If only friction in the radial coordinate is considered, the corresponding equation of motion is

$$m\ddot{r} + \frac{\partial}{\partial r} V_{\ell}(r) = F_r(r,\dot{r}) \tag{2.6.1}$$

where $V_{\ell}(r)$ is the effective potential shown in fig. 2.2.2 .

The friction force is usually assumed to be of the (classical) form

$$F_r(r,\dot{r}) = -\gamma_r \dot{r}\ f(r) \tag{2.6.2}$$

with the friction coefficient γ_r and a form factor $f(r)$ which vanishes outside the range of the nuclear interaction. Dealing with more than one coordinate q_ν one can use the Euler-Lagrange equations

$$\frac{d}{dt}\frac{\partial \mathcal{L}}{\partial \dot{q}_\nu} - \frac{\partial \mathcal{L}}{\partial q_\nu} = F_\nu(q,\dot{q}) = -\sum_\mu \gamma_{\nu\mu}\ \dot{q}_\mu\ f(q) \tag{2.6.3}$$

where $\mathcal{L} = T - V$ and F_ν denote the Lagrangian and the friction forces, respectively. The quantities $\gamma_{\nu\mu}$ are the components of the friction tensor. So far, rotational degrees and recently $\begin{bmatrix} DE\ 74 \end{bmatrix}$ vibrational degrees of freedom have been taken into account. Without going into the details of how to solve the classical equations of motion with friction forces, we mention three problems which arise in this context:

(i)   Is the friction force really of the classical form
       $F_\nu = -\sum_\mu \gamma_{\nu\mu}\ \dot{q}_\mu\ f(q)$? We shall discuss the theoretical attempts to justify this form in the following section.

(ii)   Is it possible to calculate the friction coefficient $\gamma_{\nu\mu}$? No answer has been given so far to this question.

(iii)   What are the important degrees of freedom which have to be used explicitly in the classical equations of motion ?

2.7   Microscopic derivations of friction   forces

There have been two attempts $\begin{bmatrix} BE\ 73,\ GR75a,\ HO\ 76 \end{bmatrix}$ to justify the classical form of the friction   forces given in eq. 2.6.3 . Beck and Gross $\begin{bmatrix} BE\ 73,\ GR75a \end{bmatrix}$ start out from the Liouville or von Neumann equation

$$ i\hbar \dot{\rho}(t) \;=\; \left[H, \rho(t) \right] \tag{2.7.1} $$

where $\rho(t)$ is the density matrix.   The total Hamiltonian H is split into two parts

$$ H \;=\; H_o + H_1(\vec{r}, \xi_i) \tag{2.7.2} $$

with

$$ H_o \;=\; \frac{p^2}{2m} \;+\; V(\vec{r}) \;+\; H_{intr}(\xi_i) \tag{2.7.3} $$

containing the relative part and the intrinsic part. Hence $H_1(\vec{r}, \xi_i)$ couples the relative motion to the intrinsic motion and, therefore, is responsible for friction to occur. The solution of the Liouville equation is subject to the boundary condition

$$ \lim_{t \to -\infty} \rho(t) \;\Rightarrow\; \rho_o(t) = | p_{in}(t), 0 >< p_{in}(t), 0 | \tag{2.7.4} $$

where $p_{in}(t)$ and O denote the initial wave packet and the ground states of the separated nuclei. The friction   force can now be obtained by taking time derivatives of

$$ <\vec{r}>_t \;=\; tr \left\{ \vec{r} \rho(t) \right\} \quad . \tag{2.7.5} $$

One finds

$$\frac{\partial}{\partial t} <\vec{r}>_t = \frac{1}{i\hbar} tr \left\{ \vec{r} \left[H, \varrho(t) \right] \right\} = \frac{1}{m} <\vec{p}>_t \ , \tag{2.7.6}$$

$$m \frac{\partial^2}{\partial t^2} <\vec{r}>_t = - <\vec{\nabla} V(\vec{r})>_t - <\vec{\nabla} H_1(\vec{r},\xi_i)>_t \ . \tag{2.7.7}$$

According to Ehrenfest's theorem the first term in eq. (2.7.7) is interpreted as a conservative force and the second term as the friction force

$$<\vec{F}>_t = - <\vec{\nabla} H_1(\vec{r},\xi_i)>_t = tr \left\{ \varrho(t) \vec{\nabla} H_1 \right\} \ . \tag{2.7.8}$$

This expression is evaluated in first-order perturbation theory with respect to H_1 for $\varrho(t)$ from eq. (2.7.1) . In the limit of narrow wave packets (which is necessary to justify various approximations) Beck and Gross arrive at the desired result

$$< F_\nu >_t = - \sum_\mu \gamma_{\nu\mu}(r) \ \dot{r}_\mu \tag{2.7.9}$$

which is, indeed, of the form of eq. (2.6.3) . This approach has to be criticized mainly in the use of perturbation theory of first order in calculating $\varrho(t)$ from eq. (2.7.1). The use of first-order perturbation theory implies that one is allowed to treat a certain transition completely independently from the preceding and the following transitions.

In a semi-classical formulation Hofmann and Siemens [HO 76] consider the dissipation of energy along small parts of the classical path. Around the point $\vec{r}_0$ of the path the Hamiltonian is split into two parts,

$$H = H_0(\xi_i,\vec{r}_0) + H_1(\vec{r}(t),\vec{r}_0,\xi_i) \ . \tag{2.7.10}$$

The coupling potential H_1 vanishes for $\vec{r}(t) = \vec{r}_0$ and is written as a sum of separable potentials,

$$H_1 = \sum_\alpha A_\alpha(\vec{r}(t)) \, G_\alpha(\xi_i) \, . \tag{2.7.11}$$

Now H_1 is considered as a perturbation. The unperturbed system is assumed to be in thermal equilibrium. The change of relative energy is then calculated from the expression (the brackets denote the thermodynamic average)

$$\frac{dE_{rel}}{dt} = - \sum_\alpha \frac{dA_\alpha}{dt} \langle G_\alpha \rangle \tag{2.7.12}$$

in the linear response theory. As a result one obtains a dissipative contribution which has the desired form of the Rayleigh dissipation function

$$\left(\frac{dE_{rel}}{dt}\right)_{dissipation} \approx - \sum_{\nu\mu} \gamma_{\nu\mu}(\vec{r}(t)) \, \dot{r}_\nu(t) \, \dot{r}_\mu(t) \tag{2.7.13}$$

where $\gamma_{\nu\mu}(r(t))$ is the friction tensor. It is argued that this approach is applicable to small pieces of the total path with the temperature increasing according to the dissipation of energy. As a result of this procedure the nuclei are assumed to be equilibrated in their internal motion all the way along the trajectory.

In conclusion we can state the following. There are arguments that friction forces of the type as given by eq. (2.7.9) do exist. But a completely satisfying derivation of these friction forces is not yet available.

2.8   Numerical calculations

Before we discuss some typical results of numerical calculations, let us consider the qualitative effects of friction and deformation. Fig. 2.8.1 illustrates the effect of purely radial friction on a trajectory with energy E and impact parameter b_1. For weak friction the trajectory (dashed line) is reflected from the effective potential and leads directly out of the potential again. If friction is larger (or E_{in} closer to the effective barrier $\hat{V}_{\ell_1}$, cf. fig. 2.8.1)

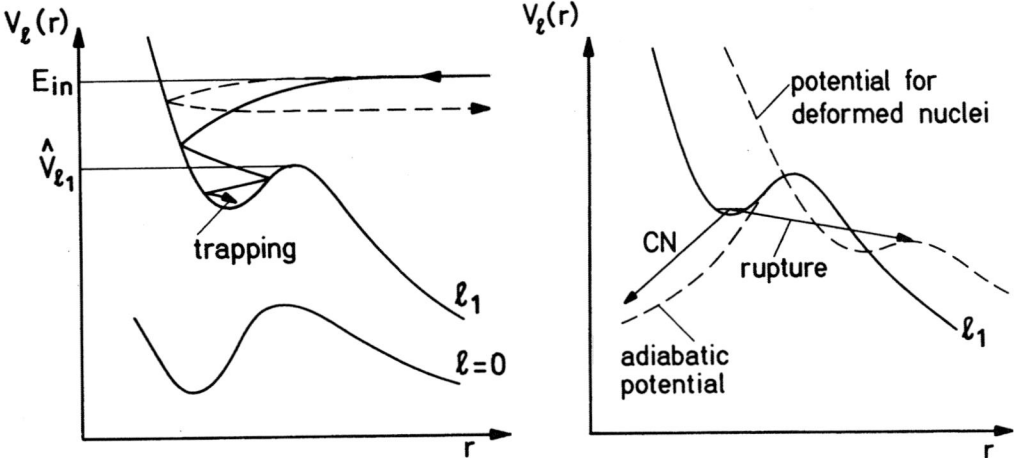

Fig. 2.8.1 Effect of radial friction
on a trajectory with E, ℓ_1.

Fig. 2.8.2 Decay possibilities of a
system after being trapped.

the system can get trapped in the pocket of the effective potential. For energies
$E_{in} < \hat{V}_{\ell_1}$ the trajectory with ℓ_1 is, of course, reflected outside the barrier.
Possible decay modes of a trapped system are shown in fig. 2.8.2 . By a readjustment
of nucleons (remember that the heavy-ion potential is assumed to correspond to a
frozen density distribution of the nuclei) the system can either form a compound-
nucleus or decay into two fragments by an enlargement of its prolate deformation .
Whereas the compound-nucleus formation is highly probable for lighter heavy-ions,
the decay through deformed states seems to be the main mode for heavier ions,
cf. [SW 72] . Figs. 2.8.3 and 2.8.4 illustrate possible effects due to the loss of
relative angular momentum (tangential friction). Due to the loss of relative angular
momentum, the effective potential is reduced during the collision. The trajectory
denoted by "without" in fig. 2.8.3 corresponds to purely radial friction leading to
the trapping of the particle. Including tangential friction and correspondingly re-
ducing the radial friction (in order to obtain the same ingoing trajectory), one
finds the trajectory denoted by "with" through which the particle gets out again once
angular momentum is reduced from its initial value ℓ_i to its final value ℓ_f . On
the other hand, tangential friction may also favour the trapping, as illustrated in
fig. 2.8.4 .

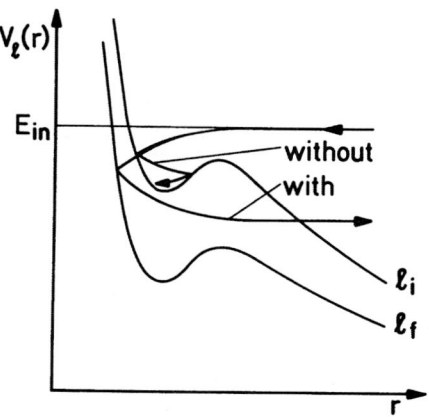

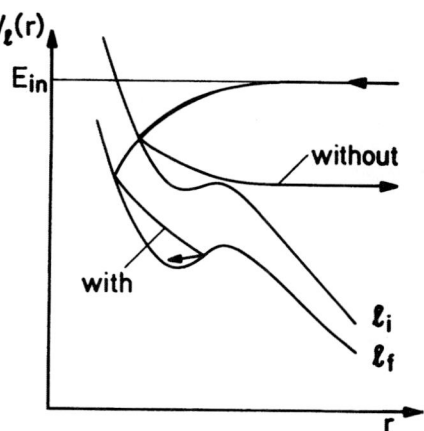

Fig. 2.8.3 Tangential friction hinder-
ing the trapping of the
system.

Fig. 2.8.4 Tangential friction favour-
ing the trapping of the
system.

Several numerical calculations of angular distributions and loss of kinetic
energy in deeply inelastic collisions have been performed during the past year
[BO 74, TS 74, GR75b , DE 75, BO 75] . We shall discuss here two of them which are
typical for the present situation of such calculations. The model of Gross et al.
[GR75b] is characterized by an unreasonably deep real potential which is obtained
from a folding procedure, cf. eq. (2.2.1) and fig. 2.8.5 . Such a deep potential

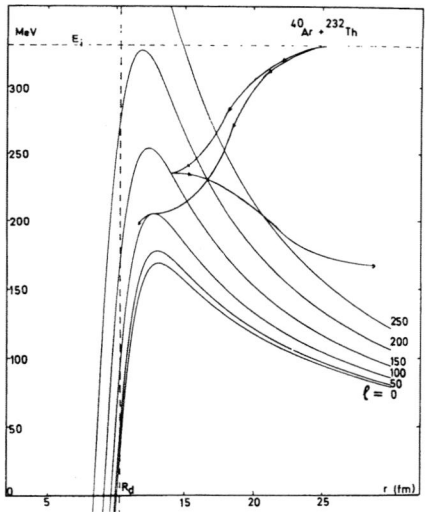

Fig. 2.8.5 Effective potential used by Gross et al. (From [GR75b]).

leads to a large increase of the relative velocity and thus, to large dissipation

of energy which increases quadratically in the velocity. Therefore, Gross et al.

obtain complete fusion whenever the trajectory comes across the barrier. In order

to get reasonable fits for the fusion cross section and the cross section for the

deeply inelastic collision, a strong long-ranged radial friction force has been

introduced. Fig. 2.8.6 shows the deflection function Θ(L) for ^{40}Ar + ^{232}Th . Due

to the strong radial friction force (tangential friction is assumed to be practically

zero) the rainbow region is stretched out over many L values giving rise to a

large cross section around the rainbow angle. The calculated cross section is com-

pared in fig. 2.8.7 with the measured cross section of K fragments. In plotting

the experimental data, the low energy part of the cross section (cf. fig. 2.5.1) is

taken at negative deflection angles [AR 73, WI 73] . The qualitative behaviour

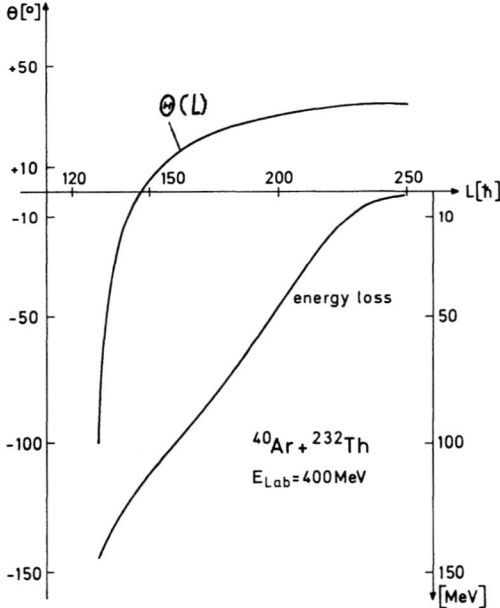

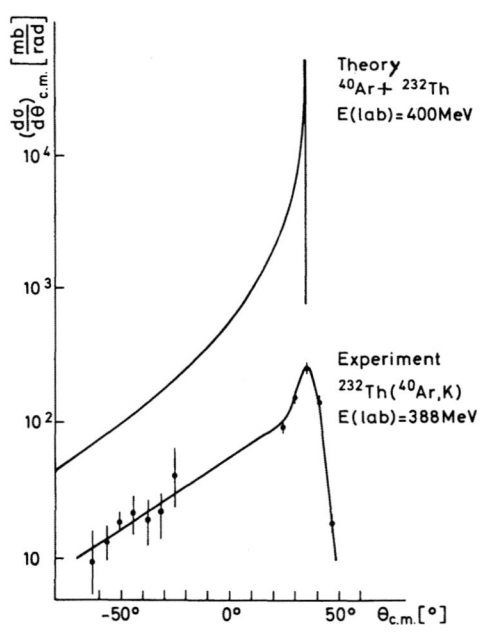

Fig. 2.8.6 Deflection angle and energy
loss as function of L for
^{40}Ar (400 MeV) + ^{232}Th.
(From [GR75b]).

Fig. 2.8.7 Calculated cross section for
the inelastic scattering of
^{40}Ar (400 MeV)+^{232}Th compared
with the experimental cross
section for ^{40}Ar (388 MeV)+
^{232}Th → K + everything.
(From [GR75b]).

of the cross section is rather well reproduced. The energy loss is shown in fig.2.8.6
as function of L and in fig. 2.8.8 as function of the deflection angle. Although
the qualitative feature of the energy loss is rather well reproduced, the calculated
values differ considerably from the experimental data. We shall see below that this
difference can be removed when the deformation of the fragments is taken into
account.

Deubler and Dietrich [DE 75] treat the scattering of two spherical
nuclei including radial as well as tangential friction. For the scattering of ^{84}Kr
(525 MeV) + ^{209}Bi they show that various qualitatively different deflection functions
can be produced with reasonable values for the real potential (with a repulsive core)
and the friction forces, cf. fig. 2.8.9 . We realize that the deflection function
turns out to be extremely sensitive to the real potential. The corresponding energies are shown in fig. 2.8.10. For spherical fragments it is impossible to obtain
an energy loss as large as required by the data. The incorporation of deformation
degrees of freedom leads to a qualitative agreement with the experimentally observed
energy loss. The importance of deformation in the exit channel has also been demonstrated by Bondorf et al. [BO 75] .

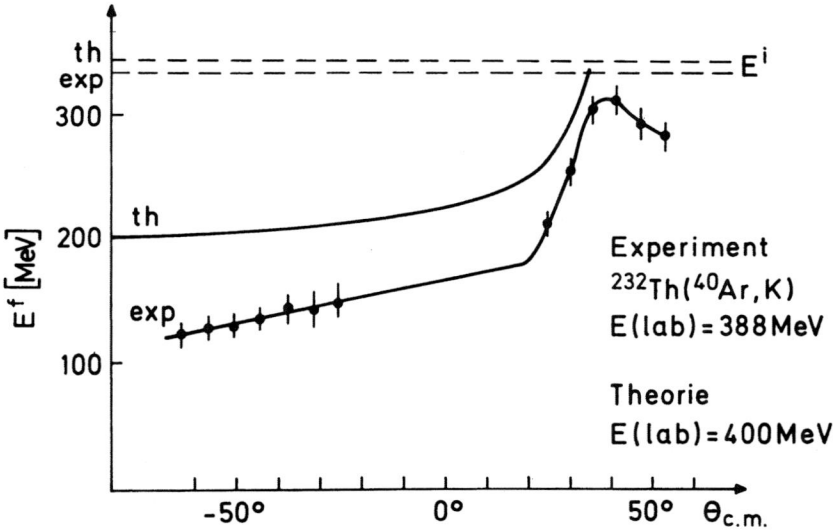

Fig. 2.8.8 Measured and calculated energy as function of the deflection
angle (From [GR75b]).

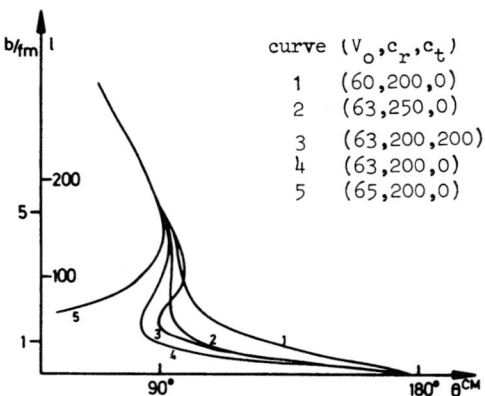

Fig. 2.8.9 Deflection function for ^{84}Kr (525 MeV) + ^{209}Bi calculated
with different values for the real potential strength
V_o (MeV), the radial and the tangential friction coef-
ficient, γ_r and γ_t (MeV fm^2 10^{-23}sec). (From [DE 75]) .

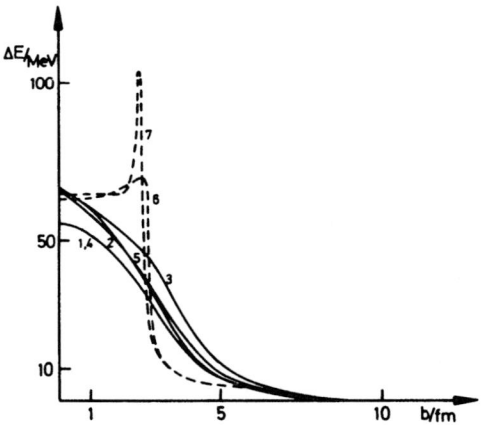

Fig. 2.8.10 Energy loss ΔE_{Kin} versus impact parameter calculated for
spherical fragments (cf. fig. 2.8.9) and deformable fragments (---).
(From [DE 75]).

2.9   Experimental cross-sections and deflection functions

As we have seen in the preceding section, it is rather difficult to get relia-
ble quantitative information about the deflection function from theoretical calculations. The reason is the high sensitivity to the parameters of such calculations
and the lack of knowledge about the precise values of these parameters. On the other
hand, the deflection functions contain important information about the reaction
mechanism. In particular, the knowledge of the deflection function could be useful
for correlating interaction times with given deflection angles. The interaction times
play a central role in the transport theory which is discussed in section 8.4 . The
problem is, can we construct deflection functions unambigously from experimental
data ?

Let us assume for simplicity that the mass transfer is negligible. Then the
cross section for inelastic processes is still determined by the deflection function,

$$\frac{d\sigma}{d\Omega} = \sum_{n} \frac{b_n}{\sin\theta} \left|\frac{db}{d\theta}\right|_{b=b_n} . \qquad (2.9.1)$$

Hence, a precise measurement of the cross section could determine the deflection
function rather unambiguously. Of course, there are several problems with this
procedure:

(i)   At present, complete and sufficiently precise information about the cross
        section is not available.
(ii) The cross section has to be decomposed into a sum over different branches of
        the deflection function.

Despite these difficulties, let us try to extract the qualitative behaviour
of the deflection functions for $^{40}Ar + ^{232}Th$, $^{63}Cu + ^{197}Au$ and $^{84}Kr + ^{209}Bi$ from
experimental cross sections. Fig. 2.9.1 is a tentative picture of the three deflection functions which are compatible with the qualitative behaviour of the deeply

inelastic cross sections discussed in section 2.5 . A continuous change from the lightest projectile ^{40}Ar to the heaviest projectile ^{84}Kr is observed. According to this picture the contribution which is denoted as fission in fig. 2.5.4 must be interpreted as the part of deeply inelastic collisions deflected to negative angles. The negative deflection angles could be due to the trapping of the system in the pockets of the effective potentials, cf. fig. 2.8.1 . Only after some time which is necessary to deform the nuclei, it is possible that the system decays, cf. fig. 2.8.2 .

It should be stressed that the deflection functions of fig. 2.9.1 are highly speculative. But it seems that the direct determination of the deflection functions from experimental data might become a useful tool in order to understand at least the kinematical part of the process.

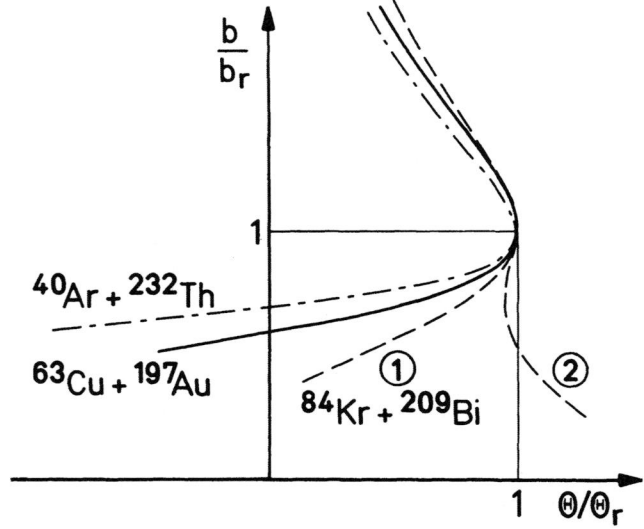

Fig. 2.9.1 Tentative picture of the qualitative behaviour of the deflection functions for ^{40}Ar (388 MeV) + ^{232}Th, ^{63}Cu(365 MeV) + ^{197}Au and ^{84}Kr (600 MeV) + ^{209}Bi .
The position of the rainbow has been taken as a reference point, in order to make the comparison between the different deflection functions easier.

2.10   Limitation due to fluctuations

In the classical scattering theory of deeply inelastic collisions which has
been considered in the previous subsections, it is assumed that the incident energy
and the impact parameter determine a well-defined trajectory. However, this may not
be entirely correct. According to the statistical picture developed in chapter 8 ,
the energy is lost in a large number of collisions which occur in a random way. This
is quite analogous to the energy loss of a Brownian particle and leads us to expect
a "spreading" of the classical trajectory. This spreading is similar to the random
walk of a particle. As a result of this random walk the one-to-one correspondence
between impact parameter b and the deflection angle Θ is destroyed. Instead of a
deflection function we encounter a deflection probability distribution $P(k_f,\Theta,b)$
where k_f is the final relative wave number. The cross section (2.9.1) takes the
more general form

$$\frac{d\sigma}{dk_f\,d\Omega} \;=\; \frac{1}{\sin\theta}\,\sum_n \int db\,b\,P_n(k_f,\Theta,b) \qquad\qquad (2.10.1)$$

where the sum is taken over all contributions leading to the same scattering angle.
A sharp distribution for $P(k_f,\Theta,b)$ leads back to the expression (2.9.1). Thus,
the suggested curves in fig. 2.9.1 can be interpreted as the ridges of the probability distribution $\int dk_f P(k_f,\Theta,b)$. It is not clear if only one ridge occurs. One could
imagine that for example two ridges (denoted by 1 and 2 in fig. 2.9.1) exist for
$b < b_r$. Then, ridge 1 should be due to the trapping and decay of the system, while
ridge 2 should be due to non-trapping.

In order to describe the effects due to fluctuations it is necessary to consider the statistical aspects of the process. Quantal corrections which are due to
interference effects, cf. section 5.1 , ãre probably strongly reduced in deeply inelastic collisions, because only averages over many channels are observed.

References

AL 75   R. Albrecht, W. Dünnweber, G. Graw, H.Ho, S.G. Steadman and J.P. Wurm,
Phys. Rev. Lett. <u>34</u> (1975) 1400

AR 73   A.G. Artukh, G.F. Gridnev, V.L. Mikheev, V.V. Volkov, J. Wilczynski,
Nucl.Phys. <u>A215</u> (1973) 91

BE 73   R. Beck, D.H.E. Gross, Phys. Lett. <u>47B</u> (1973) 143

BO 74   J.P. Bondorf, M.I. Sobel, D. Sperber, Phys. Lett. <u>C15</u> (1974) 84

BO 75   J.P. Bondorf, J.R. Huizenga, M.I. Sobel, D. Sperber, Phys. Rev.<u>C11</u>(1975)1265

BR 68   K.A. Brueckner, J.R. Buchler and M.M. Kelly, Phys. Rev. <u>173</u> (1968) 944

BR 72   R.A. Broglia, S. Landowne and A. Winther, Phys. Lett. <u>40B</u> (1972) 293

BR 74   D.M. Brink and N. Rowley, Nucl. Phys. <u>A219</u> (1974) 79

CO 75   P. Colombani, N. Frascaria, J.C. Jacmart, M. Riou, C. Stéphan, H. Doubre,
N. Poffé, J.C. Roynette, Phys. Lett. <u>55B</u> (1975) 45

DE 75   H.H. Deubler, K. Dietrich, Phys. Lett. <u>56B</u> (1975) 241

FL 74   A. Fleury, J.M. Alexander, Ann. Rev. Nucl. Sci. <u>24</u> (1974) 279

GR 74   D.H.E. Gross, H. Kalinowski, Phys. Lett. <u>48B</u> (1974) 302

GR 75a   D.H.E. Gross, Nucl. Phys. <u>A240</u> (1975) 472

GR 75b   D.H.E. Gross, H. Kalinowski, J.N. De, in ref. [HA 75] , p. 194

HA 74   F. Hanappe, M. Lefort, C. Ngô, J. Péter, B. Tamain, Phys.Rev.Lett.<u>32</u>(1974)738

HA 75   Classical and Quantum-Mechanical Aspects of Heavy-Ion Collisions, ed. by
H.L. Harney et al. (Springer Verlag,Berlin- Heidelberg-New York, 1975)

HO 76   H. Hofmann, P. Siemens, Nucl.Phys. <u>A257</u>(1976) 165

JA 75   J.C. Jacmart, P. Colombani, H. Doubre, N. Frascaria, N. Poffé, M. Riou,
J.C. Roynette, C. Stéphan, Nucl.Phys. <u>A242</u> (1975) 175

KR 75   H. Krappe, in ref. [HA 75] , p.24

LE 73   M. Lefort, C. Ngô, J. Péter, B. Tamain, Nucl.Phys. <u>A216</u> (1973) 166

LO 73   R.J. Lombard, Ann. Phys. (N.Y.) <u>77</u> (1973) 380

MC 64   J.S. McIntosh, S.C. Park and G.H. Rawitscher, Phys.Rev. <u>134B</u> (1964) 1010

MO 74   L.G. Moretto, D. Heunemann, R.C. Jared, R.C. Gatti, S.G. Thomson, in Physics
and Chemistry of Fission 1973, vol. II (IAEA, Vienna, 1974)

MY 74   W.D. Myers, same as ref. [VO 74] , p.1

NG 75   C. Ngô, B. Tamain, M. Beiner, R.J. Lombard, D. Mas and H.H. Deubler,

         Nucl.Phys. A252 (1975) 237

PE 75   J. Péter, C. Ngô, B. Tamain, J. Phys. Lett. 36 (1975) L23

RA 74   J. Randrup, W.J. Swiatecki, C.F. Tsang, Lawrence Berkeley Laboratory Report

         LBL-3603 (1974)

SC 69   W. Scheid and W. Greiner, Z. Physik 226 (1969) 364

SW 72   W.J. Swiatecki and S. Björnholm, Phys. Lett. 4C (1972) 326

TS 74   C.F. Tsang, Lawrence Berkeley Lab.Rep., LBL-2928 (1974)

VO 74   V.V. Volkov, Proc. of the Int. Conf. on Reactions between Complex Nuclei,

         Nashville, 1974, ed. by R.L.Robinson et al.(North-Holland Publ.Comp.,

         Amsterdam, 1974) vol.2, p.363

VO 75   V.V. Volkov, in ref. [HA 75] , p.254

WI 73   J. Wilczynski, Phys. Lett. 47B (1973) 484

WO 74   K.L. Wolf, J.P. Unik, J.R. Huizenga, J. Birkelund, H. Freiesleben, V.E.Viola,

         Phys. Rev. Letts. 33 (1974) 1105

3.   Gross properties of HI reactions. Compound-nucleus formation

While chapter 2 is devoted to the discussion of classical concepts in HI scattering, we now turn to a description of other features of the interaction between a pair of heavy ions. Just as in the previous chapter, we aim at an introduction of certain theoretical concepts and the description of certain gross features of the cross sections deduced from them, rather than at a detailed description of specific modes of the reaction which forms the topic of later chapters. Some of these ideas can be traced back to the work of Kaufmann and Wolfgang [KA 61] . We found it convenient to divide this chapter into two parts which are but loosely connected. In the first part, we introduce the notions of channels, Q-values, grazing collisions, critical angular momentum, the fission channel etc. The second part deals with compound-nucleus formation in HI collisions, a topic which is taken up again in chapter 8 . Here, we are only interested in limitations on compound-nucleus formation which are due to prompt fission, the existence of the Yrast line, and entrance channel effects.

3.I   Properties of reaction channels. Qualitative features of cross-sections

3.1   Channels, Q-values

We consider the reaction

$$A_1(J_1, M_1, \alpha_1) + A_2(J_2, M_2, \alpha_2) \rightarrow A_1'(J_1', M_1', \alpha_1') + A_2'(J_2', M_2', \alpha_2') .$$

Here, α_i stands for those quantum numbers which are needed besides spin J_i and magnetic quantum number M_i to specify the nuclear states uniquely, like for instance, the energy. Coupling J_1 and J_2 to the channel spin s and this with the relative orbital angular momentum ℓ_{12} to a total spin J with z-projection M , we arrive at the usual (theoretical) definition of channels c:

$$c \equiv \left\{ \alpha_1, \alpha_2, [(J_1, J_2)s, \ell_{12}] J, M \right\} . \tag{3.1.1}$$

Denoting the total energy of a fragment by E_α , and the <u>asymptotic</u> kinetic energy (large distance between the fragments) in the c.m. system by $T_{CM}^{(c)}$, we define the

Q-value of the reaction by

$$Q_{cc'} = T_{CM}^{(c')} - T_{CM}^{(c)} = E_{\alpha_1} + E_{\alpha_2} - E_{\alpha_1'} - E_{\alpha_2'} \quad . \tag{3.1.2}$$

3.2   Q-values at contact. Energy release in forming the compound nucleus

The Q-values introduced in eq. (3.1.2) are of limited significance for heavy-ion reactions, since the two ions lose a considerable fraction of their energy in overcoming the Coulomb barrier. We take this fact into account and replace in eq. (3.1.2) the <u>asymptotic</u> kinetic energy by the kinetic energy <u>at the peak of the Coulomb barrier</u>. This barrier is defined pictorially in fig. 2.2.2 , its height denoted by $V_{CB}^{(c)}$. The corresponding radius is called R_c . Different channels will in general lead to different values for barrier height and barrier radius. We denote by $T_{CM}^{(c)}$ (R_c) the kinetic energy in channel c at the barrier radius and introduce the Q-value at contact $\widetilde{Q}$ by

$$\widetilde{Q}_{cc'} = T_{CM}^{(c')}(R_{c'}) - T_{CM}^{(c)}(R_c) \quad . \tag{3.2.1}$$

Using energy conservation, we cast eq. (3.2.1) into the form

$$\begin{aligned}
\widetilde{Q}_{cc'} &= (T_{CM}^{(c')} - V_{CB}^{(c')}) - (T_{CM}^{(c)} - V_{CB}^{(c)}) \\
&= E_{\alpha_1} + E_{\alpha_2} + V_{CB}^{(c)} - E_{\alpha_1'} - E_{\alpha_2'} - V_{CB}^{(c')} \quad .
\end{aligned} \tag{3.2.2}$$

$\widetilde{Q}_{cc'}$ does not contain the angular momentum barrier and is thus appropriate only if the angular momenta and the masses in the two channels are nearly the same. To avoid misunderstandings, we emphasize that in eq. (3.2.2), $T_{CM}^{(c)}$ is the <u>asymptotic</u> kinetic energy.

To obtain an idea of the dependence of $\widetilde{Q}_{cc'}$ on mass number and charge, we consider two nuclei (Z_1, A_1) and (Z_2, A_2) in their ground states <u>at rest</u> at the distance R_c. The energy of this configuration is given by

$$E(Z_1, A_1) + E(Z_2, A_2) + V_{CB}^{(c)} \tag{3.2.3}$$

where $E(Z,A)$ is the ground state energy of the nucleus (Z,A). We compare the energy (3.2.3) with the ground state energy of the compound nucleus (Z_1+Z_2, A_1+A_2) — and thus define the energy release for formation of the compound nucleus,

$$\widetilde{E} = E(Z_1, A_1) + E(Z_2, A_2) + V_{CB}^{(c)} - E(Z_1+Z_2, A_1+A_2). \tag{3.2.4}$$

Keeping (Z_1+Z_2, A_1+A_2) fixed and considering two different fragmentations, we can calculate $\widetilde{Q}$ from the difference of the two $\widetilde{E}$ values obtained from eq. (3.2.4). We emphasize, however, that in this way only Q-values for ground-state transitions can be calculated, owing to the definition of $E(Z,A)$. However, the quantity $\widetilde{E}$ has an interest of its own because it indicates the channels open for fragmentation. We, therefore, proceed to evaluate $E(Z,A)$ with a simple model.

This is the liquid drop model. In this model, the nucleus is approximated by a spherical droplet of radius $r_o \cdot A^{1/3}$, with an energy given by [MY 66]

$$E(Z,A) = \left[1 - \varkappa\left(\frac{N-Z}{A}\right)^2\right]\left(-a_v A + a_s A^{\frac{2}{3}}\right) + \frac{3}{5}\frac{e^2 Z^2}{r_o A^{1/3}} - \frac{\pi^2}{2}\frac{e^2}{r_o}\left(\frac{\alpha}{r_o}\right)^2 \frac{Z^2}{A} \tag{3.2.5}$$

where

$$a_v = 15.677 \text{ MeV}$$
$$a_s = 18.56 \text{ MeV} \qquad r_o = 1.205 \text{ fm}$$
$$\varkappa = 1.79 \qquad \alpha = 0.546 \text{ fm} \ . \tag{3.2.6}$$

The model contains a volume energy $(-a_v A)$, a surface energy $a_s A^{2/3}$, both modified by a symmetry term $\propto\left(\frac{N-Z}{A}\right)^2$, and the Coulomb energy which contains a correction due to the surface diffuseness (parameter α). In the liquid drop model the nuclear interaction is zero for nonoverlapping nuclei. The Coulomb barrier is, therefore, completely determined by the Coulomb interaction,

$$V_{C8}^{(c)} = \frac{Z_1 Z_2 e^2}{R_c} \quad . \tag{3.2.7}$$

Here, R_c is treated as a parameter. In evaluating eq. (3.2.4), we have chosen a value for R_c given by

$$R_c = \left[0.5 + 1.36 \left(A_1^{1/3} + A_2^{1/3} \right) \right] \, fm \quad . \tag{3.2.8}$$

This is suggested by the reaction cross sections, see section 3.3 . Moreover, one chooses values for Z / A corresponding to the valley of stable masses and puts $\alpha = A_1/A$ so that $0 < \alpha < 1$. The charges are chosen so that $Z_1/A_1 = Z_2/A_2 = Z/A$, i.e. so that the charge density is kept fixed. Since for large A the ratio Z/A is appreciably less than $1/2$, this is not realistic for α close to zero or one. The dependence of E on α for various compound masses A and corresponding values of Z^2/A is shown in fig. 3.2.1 . It can be understood qualitatively by noticing that for small values of A or Z^2/A , the surface energy dominates. This energy is always increased by break-up. It favours the break-up into two fragments of very different masses over that into two equal fragments. For large values of A or Z^2/A, the relative importance of the Coulomb energy increases and finally favours the symmetric fission process over all other modes of fragmentation. For $A \approx 120$, i.e. $Z^2/A \approx 21$ the curve in fig.3.2.1 has a vanishing second derivative at $\alpha = 1/2$. For $A < 120$ the break-up into two fragments with very different masses is always energetically favoured. For $A > 120$ a critical mass asymmetry exists. For mass asymmetries more extreme than the critical value the larger nucleus tends to suck up the light one. For asymmetries less extreme than the critical the lighter nucleus tends to grow towards equality with the heavier one. For $A > 300$, i.e. $Z^2/A > 40$ the energy $\widetilde{E}$ becomes negative around $\alpha = 1/2$, i.e. the total energy of the touching fragments becomes smaller than the total energy of the compound nucleus. Note that $A \approx 300$ is the region of superheavies. We have so far neglected angular momentum. For $\ell \neq 0$, we have to include the rotational energy. For sufficiently large ℓ , this produces a dip at $\alpha = 1/2$ (or deepens it) because here the moment of inertia becomes largest.

For completeness, we mention that all nuclei with $A > 100$ are in principle unstable against fission. However, the fission barrier is so big that the lifetimes of most of these nuclei exceed the age of the universe. This instability is not visible on fig. 3.2.1 because by definition, eq. (3.2.4) , $\widetilde{E}$ is the energy release measured at the relative distance R_c . If we were to subtract from $\widetilde{E}$ the value $\frac{Z_1 Z_2 e^2}{R_c}$ of the Coulomb energy, it would become negative for $A > 100$.

The energy release $\widetilde{E}$ differs from the energy release at the scission point considered in theories of nuclear fission. The scission point is that distance in the fission process at which the two fragments are for the first time unlinked by matter. At this point they are usually deformed. This explains why such calculations [WI 64] give results which are generally similar to those shown in fig. 3.2.1 , except that the energies have smaller values, especially near $\alpha = 1/2$.

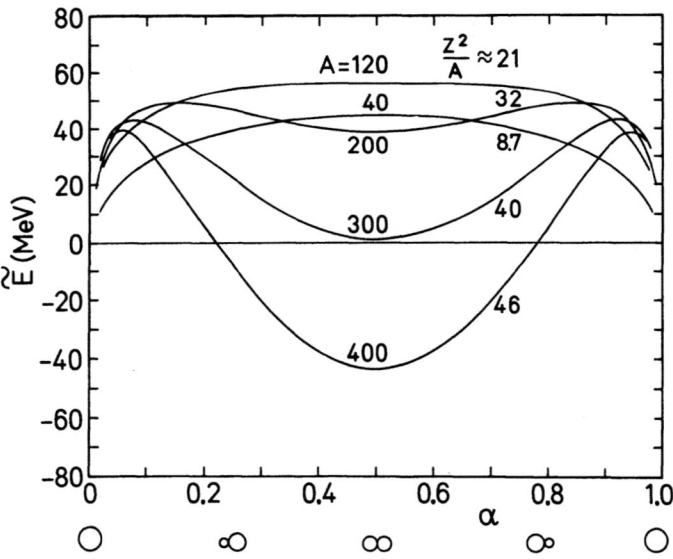

Fig. 3.2.1   The energy release $\widetilde{E}$ for formation of a compound
nucleus [SW 72] , calculated from eq. (3.2.4) along the
β-stability line $Z = 1/2$ (A−0.006·$A^{5/3}$) [BO 69] .

3.3   Grazing collisions and reaction cross-sections

Because of the strong interaction between two heavy ions, we expect that the reaction cross section σ_R can be estimated geometrically by the condition that the two ions should touch. Let R be the relative distance where nuclear reactions become important, and let b_{gr} be the impact parameter for which the distance of closest approach assumes the value R (grazing distance). Then, we expect

$$\sigma_R = \pi b_{gr}^2 = \pi R^2 \left(1 - \frac{V_{CB}}{E_{CM}} \right) \quad \text{for} \quad E_{CM} > V_{CB} \;. \tag{3.3.1}$$

The last equality (3.3.1) follows from the conservation of angular momentum which implies $b_{gr} \cdot p_\infty = R \cdot p$ where p_∞ is the momentum at infinite distance, p that at distance R . Neglecting the nuclear interaction, so that $V_{CB} = Z_1 Z_2 e^2 / R_c$, Wilczynski [WI 73] obtains a fit of $R = R_c$ to experimental reaction cross-sections as illustrated in fig. 3.3.1 with R given by eq. (3.2.8) . The interaction or grazing distance turns out to be significantly larger than that for touching spheres, for which we expect a value $1.2(A_1^{1/3} + A_2^{1/3})$ fm . In the HI potential of fig.2.2.2

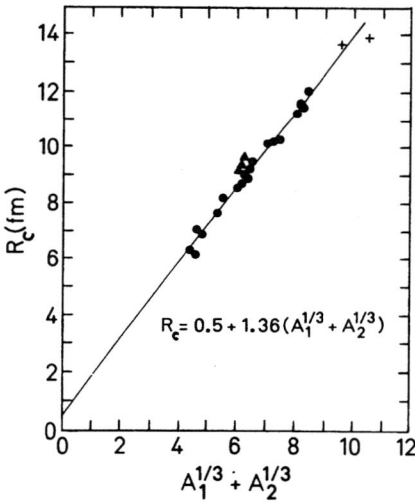

Fig. 3.3.1 Fit of σ_R to reaction cross sections by choice of R_c.
From [WI 73] . The experimental values are from [WI 63] (•),
[LE 72] (+) and [BE 71] (▲) .

V_{CB} is identified with the maximum of the barrier. Modifications are expected for deformed nuclei. Further modifications are expected, if one does not apply the simple classical picture we use here for purposes of orientation.

We introduce the grazing angular momentum $L_{gr} = p_\infty b_{gr}$ or, equivalently

$$L_{gr} = p_\infty R \left(1 - \frac{V_{CB}}{E_{CM}} \right)^{1/2} \overset{R \approx R_c}{\approx} 2\hbar\eta \frac{E_{CM}}{V_{CB}} \left(1 - \frac{V_{CB}}{E_{CM}} \right)^{1/2} \quad or$$

$$\ell_{gr} = L_{gr}/\hbar = 0.22 R \sqrt{A_{red}(E_{CM} - V_{CB})} \qquad (3.3.2)$$

where E_{CM} and V_{CB} are in MeV, R in fm. Under neglect of the nuclear interaction, the grazing angle is correspondingly defined by (cf. for example (2.4.17))

$$\theta_{gr}^c = 2 \arctan (\eta/\ell_{gr}) \approx 2\arctan \left\{ 0.72 \frac{Z_1 Z_2}{R} \left[E_{CM}(E_{CM} - V_{CB}) \right]^{-\frac{1}{2}} \right\} \qquad (3.3.3)$$

with R in fm and E_{CM}, V_{CB} in MeV. It refers to the c.m. system. Values for R and V_{CB} can be taken from fig. 3.3.2 .

It is useful to note the dependence of ℓ_{gr} on the masses of the colliding nuclei. Without Coulomb forces we simply have

$$\ell_{gr} = L_{gr}/\hbar = p_\infty R/\hbar = \sqrt{\frac{2M}{\hbar^2} E_{CM}} \sqrt{\frac{A_1(A-A_1)}{A}} R \qquad (3.3.4)$$

which steeply increases with A_1 for fixed E_{CM} . For fixed E_{Lab}

$$\ell_{gr} = \sqrt{\frac{2M}{\hbar^2} E_{lab}} \frac{\sqrt{A_1(A-A_1)}}{A} R \qquad (3.3.5)$$

increases less steeply. The Coulomb barrier V_{CB} reduces this dependence somewhat. Fig. 3.3.3 shows ℓ_{gr} calculated from eq.(3.3.2) with $R \rightarrow R_c$ for $A=40$ and $Z_1 = 1/2 A_1$,

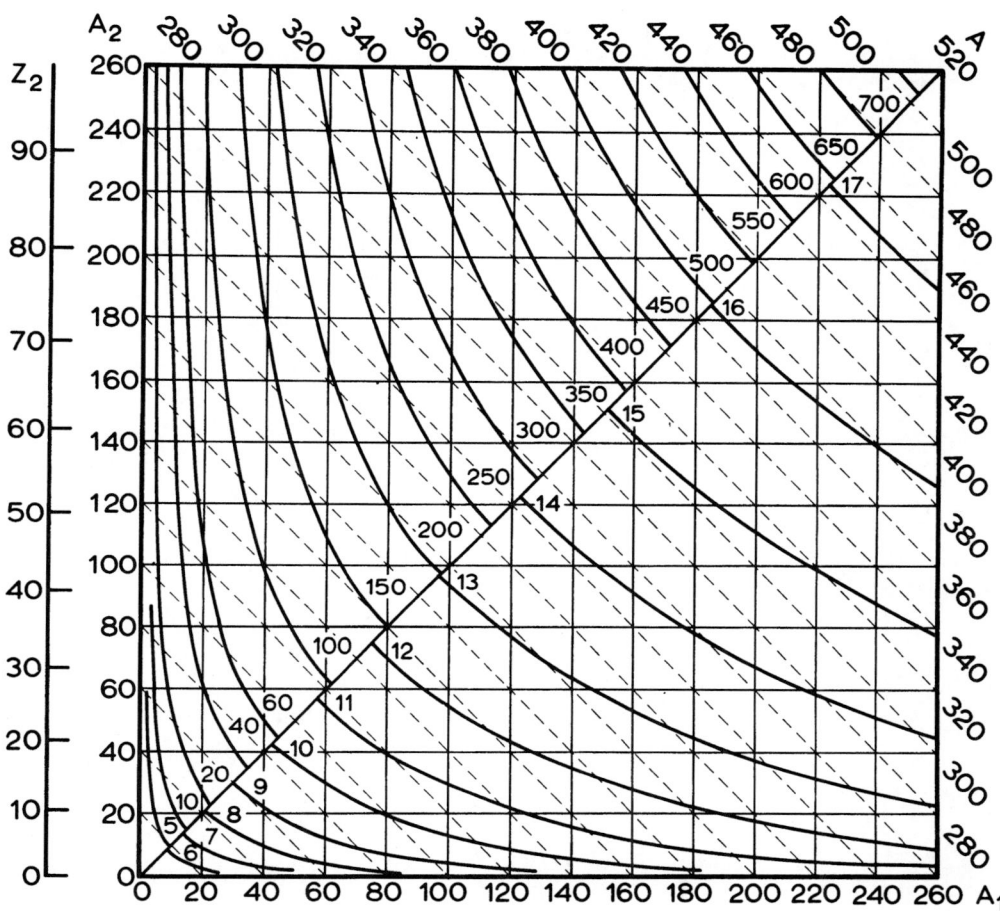

Fig. 3.3.2 Coulomb barriers $V_{CB} = Z_1 Z_2 e^2 / R_c$ in MeV (upper half)
and grazing distance $R_c = 0.5 + 1.36 \cdot (A_1^{1/3} + A_2^{1/3})$
in fm (lower half) for collisions between two β-stable
nuclei A_1 and A_2 with $Z_i = N_i - 0.006 \cdot A_i^{5/3}$.

$Z_2 = 1/2 \, A_2$. We see that for fixed E_{CM} , heavy ions carry much more angular momentum into the system than light projectiles.

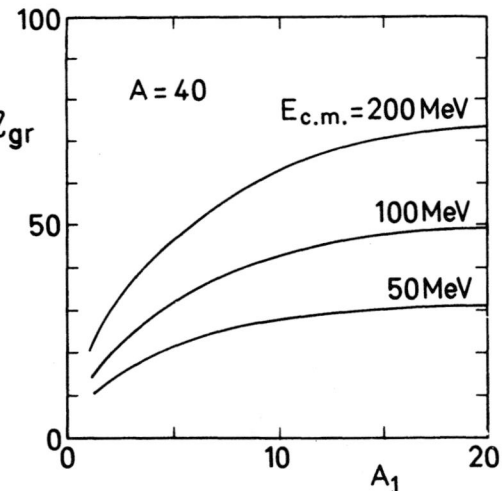

Fig. 3.3.3 Grazing angular momentum according to eq. (3.3.2)
as function of A_1 for A = 40 and different E_{CM}.

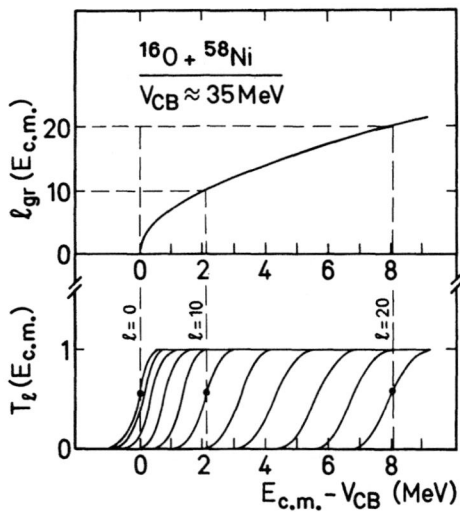

Fig. 3.3.4 Grazing angular momenta and transmission coefficients
as functions of energy.

According to eq. (2.1.9) the classical differential cross-section is given by $d\sigma/db = 2\pi b$ or with $\ell = kb = b/\lambdabar$

$$\sigma(\ell) \equiv \frac{d\sigma}{d\ell} = 2\pi \lambdabar^2 \ell \qquad (3.3.6)$$

and since all ℓ values up to ℓ_{gr} contribute to the reaction cross-section (sharp cut-off model),

$$\sigma_R = \int_0^{\ell_{gr}} d\ell \frac{d\sigma}{d\ell} = \pi \lambdabar^2 \ell_{gr}^2 \qquad (3.3.7)$$

in agreement with eq. (3.3.1). Quantum-mechanically, cf. eq. (4.1.5), ℓ has to be replaced by $\ell + \frac{1}{2}$ with ℓ integer. Hence, the integral in (3.3.7) has to be replaced by a sum yielding $(\ell_{gr}+1)^2$ instead of ℓ_{gr}^2 in eq. (3.3.7). For $\ell_{gr} \gg 1$ (a condition easily met in HI reactions) this makes no difference. In addition, the absorption is not expected to set in abruptly for $\ell \leq \ell_{gr}$, and to be zero for $\ell > \ell_{gr}$. We rather expect a smooth transition across ℓ_{gr}. We, therefore, write the reaction cross-section in the form

$$\sigma_R = \pi \lambdabar^2 \sum_{\ell=0}^{\infty} (2\ell+1) T_\ell (E_{CM}) \quad . \qquad (3.3.8)$$

We see that eq. (3.3.8) corresponds, for $\ell \gg 1$, to eq. (3.3.7) if $T_\ell (E_{CM}) = 1$ for $\ell \leq \ell_{gr}$ and zero otherwise. The typical behaviour of the "transmission coefficients" $T_\ell (E_{CM})$ as functions of angular momentum and c.m. energy is sketched in the lower part of fig. 3.3.4 , while the upper part shows ℓ_{gr} as function of E_{CM} as calculated from eqs. (3.3.2), (3.2.8) .

3.4 Qualitative decomposition of the total cross-section

We now turn to a brief discussion of the reaction mechanisms which contribute to σ_R . Qualitatively speaking (see fig. 3.4.1), we expect for $b > b_{gr}$ pure Rutherford scattering, accompanied by Coulomb excitation and, for b close to b_{gr} , nucleon transfer by tunneling processes. For $b \approx b_{gr}$ the nuclear interaction sets in. The duration time of the HI collision is expected to be short, and we, therefore, expect reactions to dominate to which compound-

nucleus formation does not contribute. These are the "direct reactions" described
in chapter 7 . For smaller values of b we encounter the "deeply inelastic collisions" discussed in chapter 2 and perhaps compound-nucleus formation. These expectations should manifest themselves both in the elastic and in the reaction cross
section. The elastic cross section has the typical behaviour shown in fig. 3.4.2
where it is plotted, divided by the Rutherford cross section σ_{Ruth} , versus
scattering angle. We note that the angle θ_{gr} at which the elastic cross section
becomes considerably smaller than the Rutherford value differs from θ_{gr}^{c} , as
defined by eq. (3.3.3). This difference is attributed to the nuclear attraction
which bends the grazing trajectory towards forward angles, as indicated in
fig. 3.4.1 .

The elastic excitation function,divided by the Rutherford value is expected to show
a similar pattern, see fig. 3.4.3 . Further details concerning figs.3.4.2 and 3.4.3
are discussed in chapter 5 .

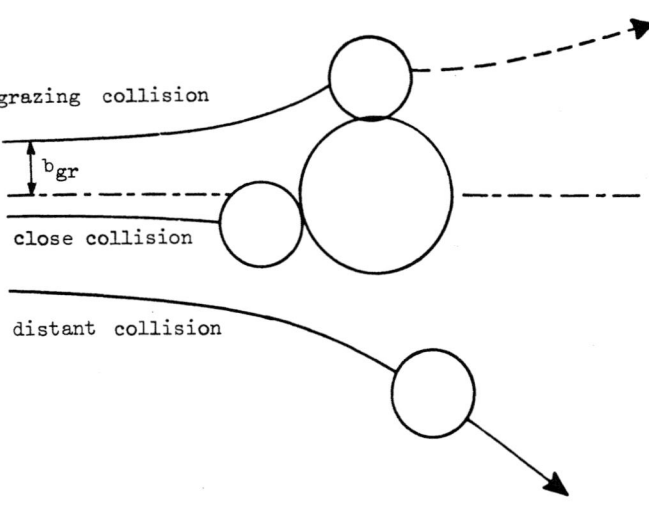

grazing collision

b_{gr}

close collision

distant collision

Fig. 3.4.1 Pictorial definition of distant collisions, grazing
collisions, and close collisions, (from [KA 61]).

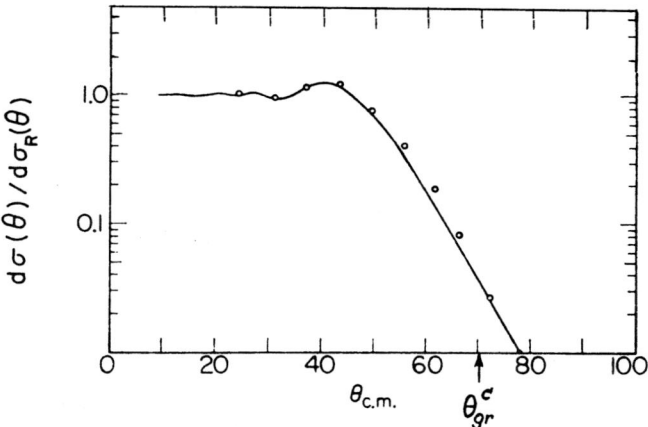

Fig. 3.4.2   Typical elastic cross section   (ratio to Rutherford) as a
function of c.m. scattering angle, for large values of Z_1
and Z_2 . The example shown is ^{16}O + ^{64}Ni at E_{lab} = 60 MeV.
(from $\left[BE\ 73\right]$) . θ^c_{gr} calculated   from eq.(3.3.3)

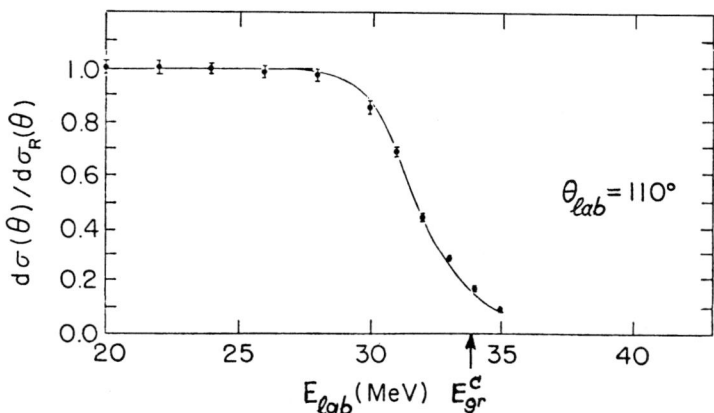

Fig. 3.4.3   Typical excitation function (ratio to Rutherford)   ,
for large values of Z_1 and Z_2. The example shown
is ^{16}O + ^{48}Ca at Θ_{lab} = 110°   (from $\left[BE\ 71\right]$) .
E^c_{gr} calculated from eq.(3.3.3)

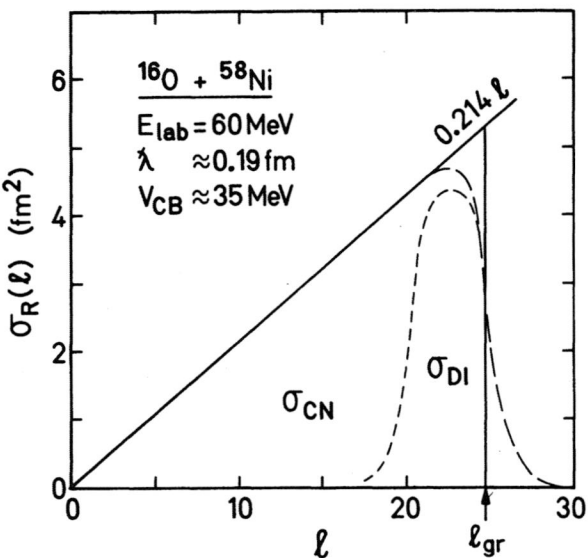

Fig. 3.4.4 Contributions to the reaction cross-section.

Concerning the reaction cross-section σ_R , the simple model of eq. (3.3.6) leads to a linear rise in ℓ , up to ℓ_{gr} , as indicated by the solid line in fig. 3.4.4 . The quantum-mechanical corrections of eq. (3.3.8) and fig. 3.3.4 lead to a smeared-out distribution as indicated by the dashed line in fig. 3.4.4 . This cross section can be decomposed into a compound part σ_{CN} , and a direct part σ_{DI} .

3.5   Critical angular momentum and instability against prompt fission

In fig. 3.4.4 we have assumed that the part of the cross section denoted by σ_{CN} leads to compound nucleus formation. This is true for ^{16}O (60 MeV) + ^{58}Ni . But for higher bombarding energies and/or heavier nuclei a third component due to deeply inelastic collisions is observed (cf. chapter 2) which is attributed to intermediate ℓ values. We discuss here one of the reasons why this component occurs: The total angular momentum in HI reactions is usually very large, see chapter 2. The high angular momentum barrier may lead to instantaneous instability of the compound system against fission. The smallest angular momentum for which this happens is called critical angular momentum and denoted by ℓ_{crit}^f . It can be estimated by using an extension of the liquid drop model introduced in section 3.2 [CO 63, SW 72, CO 74] .

Keeping the volume of the droplet fixed, one allows the droplet to undergo deformations. These deformations are described by a number of deformation parameters $(\delta_1, \delta_2, \ldots) \equiv \delta$ which are chosen in such a way that fission is geometrically describable. The total energy $E_{tot}(\delta)$ of the droplet as function of δ is given by

$$E_{tot}(\delta) = E_{tot}(0) + E_s(\delta) - E_s(0) + E_{Clb}(\delta) - E_{Clb}(0) \tag{3.5.1}$$

where $E_{tot}(0)$ is given by eq. (3.2.5), $\delta = 0$ denoting the sphere. The quantities $E_S(\delta)$ and $E_{Clb}(\delta)$ refer to the surface energy and the Coulomb energy, respectively, as functions of deformation. Fig. 3.5.1 illustrates the deformation energy $E_{tot}(\delta) - E_{tot}(0)$ as function of the quadrupole deformation δ_2. All other deformations are determined in such a way as to give the minimal energy for fixed δ_2. The different curves correspond to different nuclei which are characterized by the fissility parameter

$$x = \frac{E_{Clb}(0)}{2 E_s(0)} \approx \frac{1}{42} \frac{Z^2}{A} \approx \frac{Z}{120} . \tag{3.5.2}$$

The values of $E_{Clb}(0)$ and $E_S(0)$ are taken from eqs. (3.2.5) and (3.2.6). The last near equalities hold near the β-stability line for large Z. It turns out that for $x = 1$, the droplet is instable against quadrupole deformations, so that $x > 1$

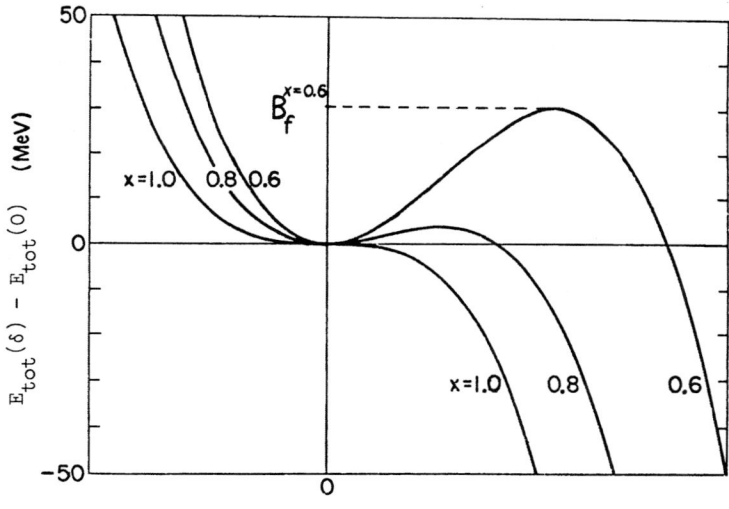

Fig. 3.5.1 The liquid drop energy versus quadrupole deformation.
(From [NI 72]).

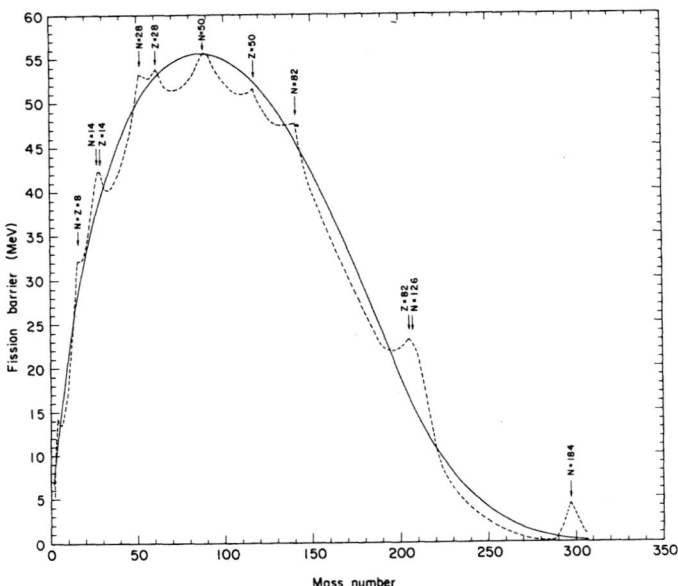

Fig. 3.5.2 The height of the fission barrier versus mass number. The dashed line corresponds to the inclusion of shell corrections in the mass formula. (From [MY 66]).

marks nuclei for which the fission barrier vanishes. The value x = 1 is realized for $Z \approx 120$. The height of the fission barrier is denoted by B_f in fig. 3.5.1 . For sufficiently large deformation, the droplet separates into two droplets. Within this model of a deformable droplet of constant volume, one can calculate the height of the fission barrier versus mass number A , for masses in the valley of stable nuclei. The result is shown in fig. 3.5.2 .

We have so far considered a system with total angular momentum zero and turn now to cases where $\ell > 0$ [SW 72, CO 74] . We add to $E_{tot}(\delta)$ the rotational energy, so that

$$E_{tot}(\delta, \ell) = E_{tot}(0) + E_s(\delta) - E_s(0) + E_{C\ell b}(\delta) - E_{C\ell b}(0)$$
$$+ E_{rot}(\delta, \ell) \quad . \tag{3.5.3}$$

For $E_{rot}(\delta, \ell)$ we take the expression $\hbar^2 \ell(\ell+1)/[2 \mathcal{F}_{rigid}(\delta)]$, where $\mathcal{F}_{rigid}(\delta)$ is the moment of inertia of a rigid body with deformation δ . For fixed ℓ , one can ask for a (local) minimum of E_{tot} as a function of δ . For $\ell = 0$, this minimum is at $\delta = 0$, while for $\ell > 0$, nonspherical equilibrium shapes are attained. The energy E_{tot} depends on two dimensionless parameters, x (see eq. (3.5.2)) and

$$y = \frac{E_{rot}(\delta = 0, \ell)}{E_S(\delta = 0)} \approx 2.0 \frac{\ell^2}{A^{7/3}} \qquad . \qquad (3.5.4)$$

The factor $A^{7/3}$ in eq. (3.5.4) is easily understood. The surface energy contributes $A^{2/3}$, and the moment of inertia $A^{5/3}$. In discussing the dependence of E_{tot} (x,y), we have to keep in mind that it is an idealization to replace the actual nucleus by a liquid drop (shell effects!), and to take the rigid body value for the moment of inertia. Qualitatively speaking, the addition of angular momentum enhances the drop-let's ability to undergo fission, since $E_{rot}(\delta,\ell)$ decreases with increasing defor-mation. Some quantitative results, taken from $\left[\text{SW 72, CO 74}\right]$, are shown in fig.3.5.3. For small rotational energy (ℓ small, y small), the equilibrium shapes are axially symmetric and correspond to flat pancakelike structures. For larger ℓ-values (above the first solid line in fig. 3.5.3) triaxial shapes become stable at equilibrium, but there still exists a barrier against fission. This barrier is gradually reduced with increasing ℓ, until it vanishes (second solid line in fig. 3.5.3). Above this

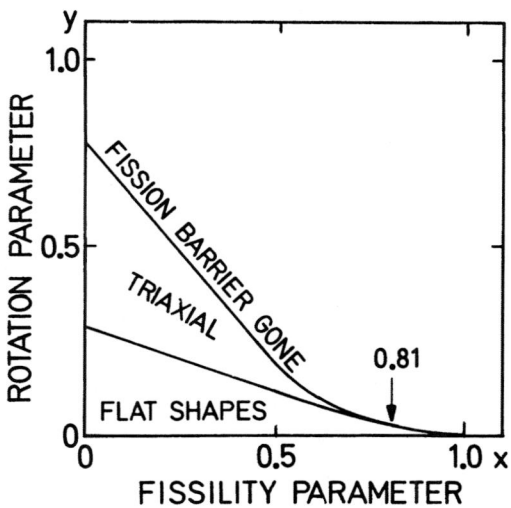

Fig. 3.5.3 Equilibrium shapes for rotating systems. (From $\left[\text{SW 72}\right]$).

line, the nucleus is instable against prompt fission, and compound-nucleus formation is impossible. The two lines merge at x=0.81 . The second solid line in fig. 3.5.3 defines, for each value of x (or of A) the <u>critical angular momentum</u> ℓ_{crit}^{f} (A) . This function, deduced from fig. 3.5.3 , is shown in fig. 3.5.4 . The physical significance of ℓ_{crit}^{f} is quite different from that of the quantity ℓ_{crit} introduced in section 2.3.3 . Nevertheless, both quantities are referred to as "critical angular momentum" in the literature.

In using fig. 3.5.4 one should keep in mind that it cannot be expected to give quantitative agreement with the data, because of shell effects, because $\mathcal{F}_{rigid}$ may not be the correct moment of inertia, and because we know little about nuclear dynamics for high values of ℓ .

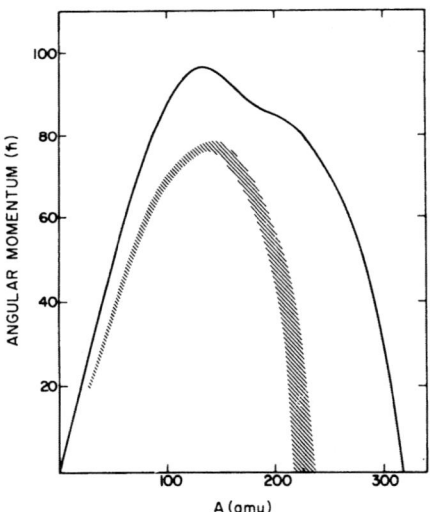

Fig. 3.5.4 The critical angular momentum ℓ_{crit}^{f} as function of mass number. The solid line gives the value of angular momentum at which the fission barrier of β-stable nuclei of mass number A is predicted to vanish. The hatched area indicates the region of competition between fission and particle emission, cf. section 3.6. (From [PL 74a]).

3.6   Prompt fission and compound-nucleus formation

The existence of a critical angular momentum ℓ^f_{crit} limits compound-nucleus formation in HI reactions. This can be seen from fig. 3.6.1 which classifies nuclear reactions as functions of the collision parameter b and the c.m. energy E_{CM} . Recall that $\sigma_R = \pi b^2_{gr}$, cf. eq. (3.3.1). The figure shows three areas, denoted by "distant collisions", "grazing collisions", and "close collisions", respectively, in an obvious fashion. (See section 3.4). The critical angular momentum ℓ^f_{crit}, defined by the condition $B_f = 0$, corresponds to the dashed line in fig. 3.6.1 . This is because $\ell \propto b \sqrt{E_{CM}}$, so that the lines of constant ℓ values are hyperbolas. To the right and above the dashed line, compound-nucleus formation is impossible, and we expect the ratio σ_{CN}/σ_R to decrease with increasing E_{CM} , where σ_{CN} denotes the cross section for compound-nucleus formation. A measurement of this ratio gives an indication of the value of ℓ^f_{crit} . Data relating to this question are discussed in section 3.10 below.

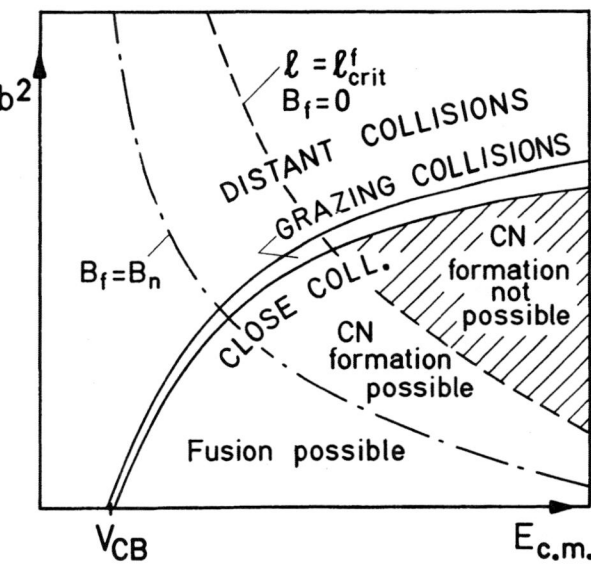

Fig. 3.6.1 A classification of nuclear collisions in the b^2 vs. E_{CM} plane, and limitations arising from ℓ^f_{crit} . (From [SW 72]).

The dot-and-dash line in fig. 3.6.1 corresponds to the condition where $B_f = B_n$, the separation energy for the last neutron. To the left and below this line, we have $B_f > B_n$, so that the phase space available for neutron emission exceeds the density of states at the fission barrier. In this region, we expect the compound nucleus, once it is formed, to decay by emission of light particles and, therefore, to survive (fusion). Between the two curves labelled $B_f = B_n$ and ℓ_{crit}^f, ($B_f = 0$) on the other hand, the compound nucleus should decay preferentially through fission.

Within the sharp cut-off model introduced in section 3.3 the compound-nucleus formation cross section is given by

$$\sigma_{CN} = \pi \lambda^2 \sum_{\ell=0}^{min\,(\ell_{crit}^f,\,\ell_{gr})} (2\ell+1) = \begin{cases} \sigma_R & for \ \ell_{gr} < \ell_{crit}^f \\ \pi \lambda^2 \,(\ell_{crit}^f + 1)^2 & for \ \ell_{gr} > \ell_{crit}^f \end{cases} \qquad (3.6.1)$$

if one assumes that a compound nucleus is formed whenever this is possible.

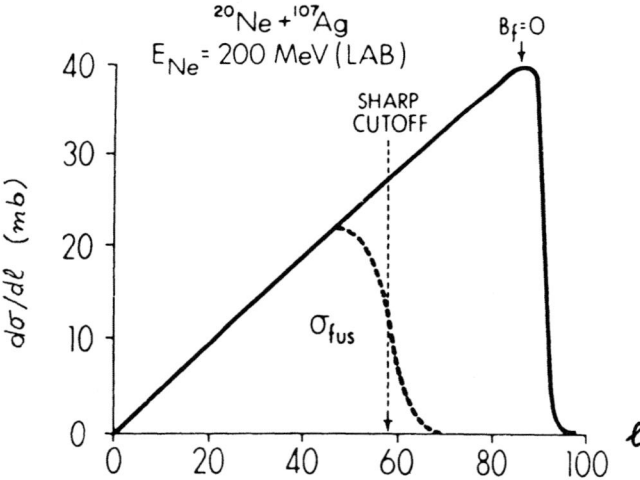

Fig. 3.6.2 Calculated fusion cross section as a function of angular momentum. The dashed curve represents the cut-off imposed by fission as calculated by Blann and Plasil [BL 72a] using the Bohr-Wheeler [BO 39] expressions for neutron evaporation and fission widths, modified by including proton and alpha-particle emission. (From [BL 72a]).

The fusion cross section σ_{fus} is smaller than σ_{CN} from eq. (3.6.1) because of the large fission probability for $B_f < B_n$, see fig. 3.6.2 . These considerations are only semiquantitative, and we have only given necessary (not sufficient) conditions for compound-nucleus formation and fusion.

3.II. Level densities and the compound nucleus

For mass numbers $A < 54$, the binding energy per nucleon increases with A . In this domain, compound nucleus formation by HI reactions leads to fairly high excitation energies. For $A > 54$, the binding energy decreases with A . The Coulomb barrier increases, however, so that significant compound-nucleus formation in HI reactions is again related to high excitation energies of the compound nucleus. Moreover, the angular momentum of a HI reaction is often very high. We now discuss compound-nucleus formation in the light of these facts. We ask for the density of compound-nuclear levels available at a certain excitation energy E^* and with a certain spin J and parity π . We take this density $\rho(E^*, J^\pi)$ as a preliminary measure for the probability of compound-nucleus formation. (Similarly, the density of final states in the fragments strongly influences the decay width of the compound nucleus.) We, therefore, work out $\rho(E^*, J^\pi)$. A review of level densities has been given by Bloch $\begin{bmatrix} BL\ 72b \end{bmatrix}$ and Huizenga $\begin{bmatrix} HU\ 72 \end{bmatrix}$. In $\begin{bmatrix} BO\ 69 \end{bmatrix}$ Appendix 2B the derivation of the fundamental level-density formulas is given.

3.7   Nuclear level densities

Let us consider the nucleus in the independent particle (or quasi-particle) model, where the ground state is given by filling the lowest single-particle states according to the Pauli principle up to the Fermi energy ε_F (see fig. 3.7.1) . All residual interactions which can shift levels are neglected. The procedure of finding the nuclear level density is to count the number of different configurations corresponding to a given excitation energy E^*, spin J and parity π . It turns out that the level density $\rho(E^*, J^\pi)$ depends only upon the density D_o^{-1} of single-particle states near the Fermi energy ε_F , all other properties of the single-particle potential being irrelevant. The reason is that (for fixed E^* and J^π)

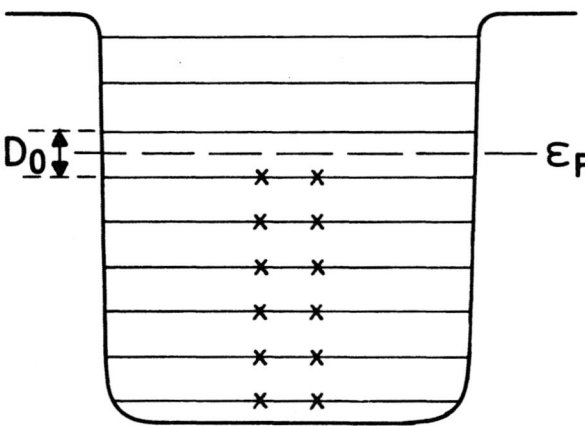

Fig. 3.7.1 Shell model as a basis for obtaining the level density.

the overwhelming number of configurations comes from the excitation of a large number of nucleons into states right above the Fermi energy. The number of configurations where one or a few nucleons take up the entire excitation energy E^* is comparatively small.

The relevant information on the single-particle potential is contained in the density

$$g(\varepsilon) = \sum_i \delta(\varepsilon - \varepsilon_i) \qquad (3.7.1)$$

of single-particle levels $\varepsilon_i > 0$, where each state is to be counted according to its multiplicity, i.e. $i \equiv (n_i, \ell_i, j_i, m_i)$ is the collection of all single-particle quantum numbers. The level density $\tilde{\rho}(E^*, M)$ at the energy E^* and for the z-projection M of the total spin is then given by the number N of solutions (with total energy E^* within a chosen unit energy interval) of the three equations

$$A = \sum_i \nu_i \quad, \quad E^* = \sum_i \nu_i \varepsilon_i - E_{ground\ state} \quad, \quad M = \sum_i \nu_i m_i \qquad (3.7.2)$$

with $\nu_i = 0$ or 1 . The level density $\rho(E^*,J)$ for the total angular momentum J is finally determined from

$$\rho(E^*,J) = \widetilde{\rho}(E^*,M=J) - \widetilde{\rho}(E^*,M=J+1) \tag{3.7.3}$$

which follows from $\widetilde{\rho}(E^*,M) = \sum\limits_{J \geq M} \rho(E^*,J)$. For completeness we note that in order to find $\rho(E^*,J^\pi)$ one has to consider a fourth equation in (3.7.2) which fixes the total parity π of the configurations. In general the two parities are expected to be equally distributed, i.e. $\rho(E^*,J^\pi) = \frac{1}{2} \rho(E^*,J)$. We do not discuss here how the combinatorial problem of finding N is solved. We only give some of the most frequently used formulas for $\rho(E^*,J)$ and describe the approximations under which they have been obtained. These approximations are introduced because a closed expression for $\rho(E^*,J)$ in terms of ε_i does not exist.

The most common approximation is the "continuous approximation" where $g(\varepsilon)$ is replaced by a continuous function $\overline{g}(\varepsilon)$. This approximation is valid only if $E^* \gg \{\overline{g}(\varepsilon_F)\}^{-1}$ which is the mean single-particle level distance D_0 at the Fermi energy. Since D_0 ranges from 0.1 to 1 MeV only, $E^* \gg D_0$ is a rather weak condition. The actual calculation of $\rho(E^*,J)$ is done using a Laplace transform and evaluating it,after a series expansion, with the saddle-point method [BO 69, appendix 2B] . Since one neglects in the saddle-point method the energy dependence of $\overline{g}(\varepsilon)$ the result is valid only for not too large excitations, i.e. $E^* \ll \varepsilon_F A^{1/3}$. One then finds [BO 71]

$$\rho(E^*,J) = \frac{\sqrt{a}}{24} \left(\frac{\hbar^2}{2\,\mathcal{F}_{rigid}} \right)^{3/2} \frac{(2J+1)\exp\left\{2\sqrt{a\left(E^* - \frac{\hbar^2}{2\,\mathcal{F}_{rigid}} J(J+1)\right)}\right\}}{\left[E^* - \frac{\hbar^2}{2\,\mathcal{F}_{rigid}} J(J+1) \right]^2} \tag{3.7.4}$$

where $a = (\pi^2/6)\overline{g}(\varepsilon_F)$, and $\mathcal{F}_{rigid} = \frac{\hbar^2}{2} \frac{A^{5/3}}{35} \, MeV^{-1}$ is the rigid-body moment of inertia. In eq. (3.7.4) neutrons and protons have been treated separately. Different approximations in the evaluation of $\rho(E^*,J)$ often lead to different E^* dependences in the denominator of eq. (3.7.4). This has little effect since the expression is dominated by the exponential. The appearance of the rigid-body moment of inertia is unphysical and related to the approximations. Improvements [RO 57] have been attempted which show that $\mathscr{F}_{rigid}$ should be replaced by $\alpha\,\mathscr{F}_{rigid}$ with α between 1/2 and 1 . Eq.(3.7.4) shows that for fixed E^*, $\rho(E^*_{}J)$ increases with increasing J until J reaches the value $\approx (2\,\mathscr{F}_{rigid}E^*\!/\hbar^2)^{1/2}$. For these values of $J(E^*)$, connected with the Yrast line discussed below, the expression (3.7.4) becomes very small and actually has to be replaced by a better approximation . If E^* is sufficiently far from this value, we can expand the square root in eq. (3.7.4) and neglect the J dependent term in the denominator. This gives the more familiar expression

$$\rho\,(E^*_{}\mathscr{J}) = \frac{\sqrt{a}}{24}\left(\frac{\hbar^2}{2\,\mathscr{F}_{rigid}}\right)^{3/2}\frac{exp\,(2\sqrt{aE^*})}{E^{*2}}\,(2\mathscr{J}+1)\,exp\left[-\frac{\mathscr{J}(\mathscr{J}+1)}{2\sigma^2}\right] \qquad (3.7.5)$$

where the "spin cut-off factor" σ is determined by

$$\sigma^2 = \sqrt{\frac{E^*}{a}}\Big/\left(\frac{\hbar^2}{\mathscr{F}_{rigid}}\right) = \frac{'nuclear\ temperature}{\hbar^2/\mathscr{F}_{rigid}} \qquad . \qquad (3.7.6)$$

Modifications have been introduced mainly to account for closed shells, for pairing forces, and for deformations. For a system with shell structure the continuous approximation for $g(\varepsilon)$ is not valid. Then it is advisable to replace E^* by $E^*_R = E^* + R$ where R is the Rosenzweig correction [RO 57] . To include the pair correlations which may lower the ground-state energy and thus change E^* , it is customary to replace E^* by $E^*_P = E^* - \delta\Delta$ where Δ is the pairing energy, and where $\delta = 2,1,0$ for even-even, odd-A and odd-odd nuclei, respectively. To within a good approximation Δ is given by $\Delta = 16.2\ MeV/A^{0.551}$ [TH 68] or $\Delta = 12\ MeV\ A^{-1/2}$ [BO 69] . Deformations of the well, important for deformed nuclei, have been incorporated into the level-density formula [KA 66, BJ 74] . Isospin can also explicitly be taken into account [KA 68, BL 72b] . There are also methods to take into account the residual interaction left out in eq. (3.7.4) [CH 71, FR 72] .

The parameters of the model to be determined by fits to the data are a, $\mathscr{F} = \alpha\,\mathscr{F}_{rigid}$ and Δ . In the fits to the data, Δ incorporates both pairing and shell effects. Information on level densities is available from the following sources [TH 68, HU 72] : (i) Counting of levels observed in slow neutron scattering experiments. This method works for practically all A and gives information on $\rho(B_n, J)$ where B_n is the neutron binding energy, and J is restricted by the orbital angular momentum 0 or 1 of the incident neutron. A careful analysis is required to obtain unique spin and parity assignments. (ii) Counting of resonances observed in reactions with charged particles. Analogue resonances in medium-weight nuclei are particularly useful examples because here all enhanced fine-structure resonances have the same spin and parity. (iii) Spectroscopy of outgoing particles in a reaction to determine the level density in the residual nucleus. This method is restricted to fairly low excitation energies, due to resolution problems. (iv) Use of the statistical theory of nuclear reactions to determine level densities from evaporation spectra. (The statistical model is discussed in section 8.1). (v) Use of absolute cross sections to determine Γ/D from Hauser-Feshbach fits and, subsequently, Γ from cross section fluctuations or from channeling. Here, Γ is the average width of the compound nucleus and $D = \rho^{-1}(\overset{*}{E})$. Methods (iv) and (v) are useful also in the region of overlapping resonances $(\Gamma > D)$ where the counting methods (i) to (iii) cannot be applied. A good overall value for a seems to be $a = A/8$ MeV^{-1} . Values for Δ have to be given individually for each nucleus [TH 68, HU 72] . The spin cut-off factor σ is less than the value given by a rigid-body moment of inertia, at least for heavy nuclei [HU 72] .

3.8   Yrast levels [TH 68, JO 73]

        For fixed J there are no levels below some lowest (yrast) energy Y_J . The yrast line connecting all these points in the E^* vs. J plane is qualitatively shown in fig. 3.8.1 . Beringer and Knox [BE 61] have obtained Y_J , see fig. 3.8.2 , from the liquid drop model by calculating the minimum total energy for fixed J from the expression (3.5.3) allowing only for spheroidal shapes. For $A \geq 100$, the cases in fig. 3.8.2 correspond very closely to $Y_J = \left[\hbar^2/(2\mathcal{I}_{rigid})\right]J(J+1)$. This work has been extended to non-spheroidal shapes by Cohen et al. [CO 63, CO 74] , where also the critical angular momentum, corresponding to prompt-fission instability has been

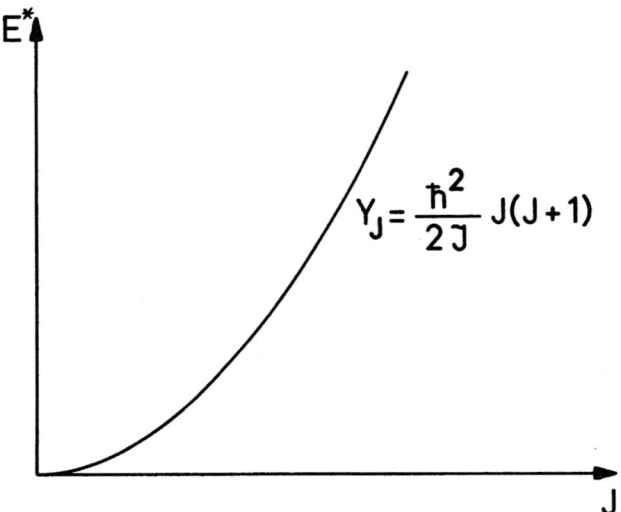

Fig. 3.8.1 Yrast line (schematic).

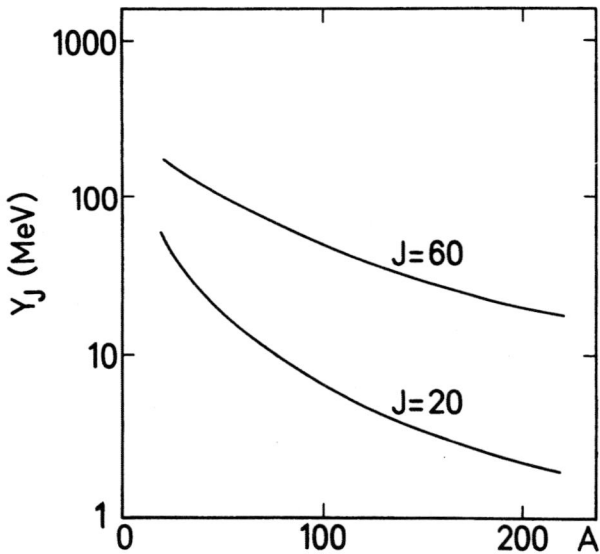

Fig. 3.8.2 Yrast energy Y_J as function of mass number for various
J as calculated from the liquid drop model [BE 61] .

discussed (see section 3.5.) . Calculations of yrast lines have also been done in the
independent-particle model plus pairing [GR 67] , for a Fermi gas in a square well
or a harmonic oscillator well [SP 63] and in the statistical model [LA 63] .

All these models lead essentially to

$$Y_{\mathcal{J}} = \frac{\hbar^2}{2\mathcal{F}} \; \mathcal{J}(\mathcal{J}+1)$$

(3.8.1)

with 1/2 $\mathcal{F}_{rigid} \lesssim \mathcal{F} \lesssim \mathcal{F}_{rigid}$. Summarizing we may say that although we have
some idea of the yrast levels from the calculations, we do not know the yrast
energies too precisely.

Experimentally [JO 73] our knowledge of the yrast levels is limited to the
region J $\lesssim$ 20 : Mostly through (α,4n) reactions, high spin states are populated
which decay subsequently by γ-emission down to the yrast line and then down along
the yrast line, see fig. 3.8.3. The failure to observe higher J-values (there is
only a broad γ-continuum which cannot be resolved) is ascribed to the existence of
several bands with nearly the same spacings that lie close to the yrast line.

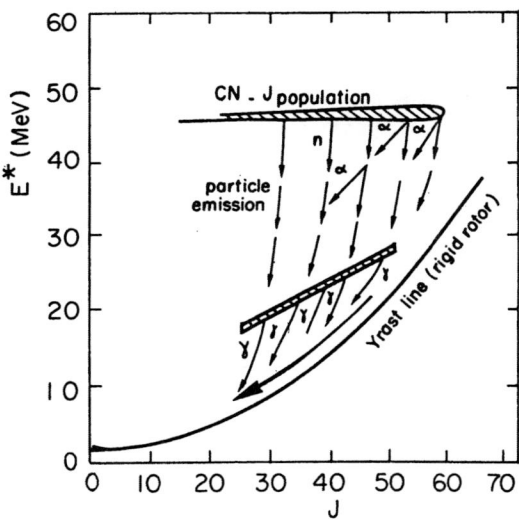

Fig. 3.8.3 Population of yrast levels in (α,xn) or HI reactions.
(From [LE 75b]).

More recently also ions like ^{11}B, ^{14}N, ... and also ^{40}Ar have been used particularly at Berkeley to populate high spin states in the compound nucleus and to measure the subsequent γ-decay. The rotational band shows very significant deviations from the simple J(J+1) rule. A typical example is given in fig. 3.8.4 . These deviations are due to changes in the moment of inertia. Since these changes are not yet completely understood, the extrapolation to higher J values is difficult.

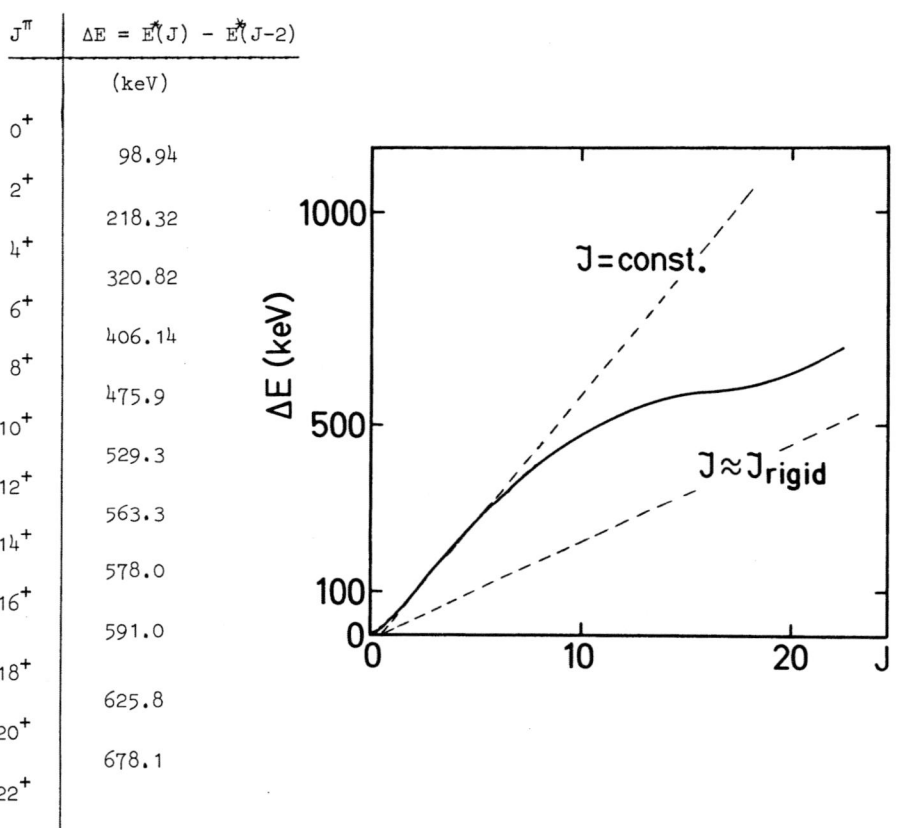

J^{π}	$\Delta E = E^*(J) - E^*(J-2)$
	(keV)
0^+	
	98.94
2^+	
	218.32
4^+	
	320.82
6^+	
	406.14
8^+	
	475.9
10^+	
	529.3
12^+	
	563.3
14^+	
	578.0
16^+	
	591.0
18^+	
	625.8
20^+	
	678.1
22^+	

Fig. 3.8.4 Rotational band of ^{158}Dy . The lines were assigned by
γ-γ coincidence techniques. (From [JO 73]).

3.9   Yrast levels and compound-nucleus formation

The limitation for compound-nucleus formation due to the instability against prompt fission has been discussed in section 3.6. Let us discuss now the consequence of level densities on the compound-nucleus formation in the E^* vs. J diagram of fig. 3.9.1 . We have indicated the yrast line by the solid curve. To the left of this line,levels are available in the compound nucleus. A second line, indicated as population line in fig. 3.9.1 (dot and dash), is the boundary in the E^* vs. J plane given by the angular momentum of the incoming ions. Only the area to the left of and above the population line is available in the HI reactions. $\widetilde{E}$ is the energy release under compound-nucleus formation from the channel containing the two ions, as defined in eq. (3.2.4) . As is seen from fig. 3.2.1, $\widetilde{E}$ is usually positive. The population line is given by

$$E_{pop} = \widetilde{E} + \frac{\hbar^2}{2MA_{red}R^2}\ J(J+1) \tag{3.9.1}$$

where the last term represents the orbital energy for the grazing collision (we assume the spins of the nuclei to be negligible).The excitation energy E^* for a certain HI collision is fixed by the sum of the c.m. energy of the colliding ions and the Q-value for the formation of the compound nucleus, $E^* = E_{CM} - V_{CB} + \widetilde{E}$.

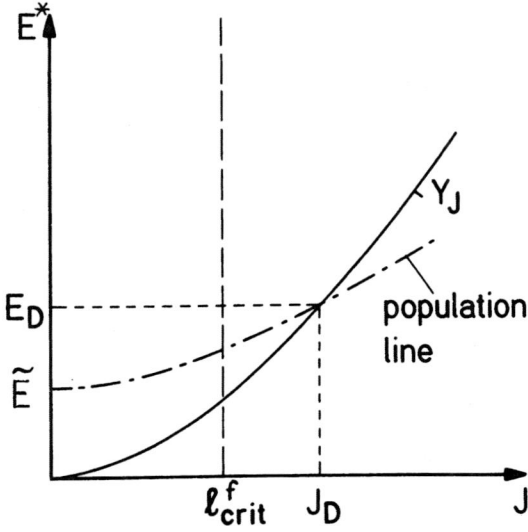

Fig. 3.9.1 Limitations in compound-nucleus formation due to the yrast line.

For $\widetilde{E} > 0$ the crossing point $E_D(J_D)$ indicates the energy E^* above which the
compound-nucleus cross-section σ_{CN} has to be less than the reaction cross-section
because there are for fixed $E^* > E_D$ no compound levels between the yrast line and
the population line. Values for E_D and J_D can be calculated from eqs. (3.2.4),
(3.8.1) and (3.9.1) with $\mathcal{F} = \mathcal{F}_{rigid}$ and R the grazing distance given by
eq. (3.2.8). One finds that J_D is usually considerably larger than ℓ_{crit}^f (see
section 3.5). Therefore, the limitation of compound-nucleus formation is usually
due to fission and not to the absence of compound levels according to the yrast line.
For $\widetilde{E} < 0$ the compound nucleus is energetically instable against breakup into two
touching fragments, cf. eq. (3.2.4). According to fig. 3.2.1 this happens only for
A > 300, and for nearly symmetric fragmentations. Since compound-nucleus levels are
available only for $E^* > 0$, the c.m. energy of the colliding ions has to be larger
than the Coulomb barrier V_{CB} by at least $|\widetilde{E}|$ in order that compound-nucleus
formation becomes possible.

We have not touched upon the question whether or not the compound nucleus is
actually formed. In addition to the necessary conditions, (i) that compound levels
be available and (ii) that a barrier against fission must exist, there is the
question whether the constituent nucleons of the two colliding nuclei rearrange
in such a way that compound-nucleus formation actually takes place. This dynamical
problem is not yet understood. It is connected with the problem of friction or vis-
cosity, i.e. the transfer of energy from one degree of freedom (the relative motion)
to the many degrees of freedom describing the compound nucleus. Fig. 3.9.1 as well
as the corresponding fig. 3.6.1 only depict those areas where compound-nucleus
formation is either possible or altogether impossible.

3.10   Experimental results on fusion and compound-nucleus formation
        Entrance channel effects

The compound nucleus is defined as the totally equilibrized intermediate
stage in a nuclear reaction, with a lifetime large compared to the duration of a
direct HI reaction. The highly excited compound nucleus can decay either by emission

of light particles and γ-rays, or by (delayed) fission[†]. The first alternative is
called fusion and applies mostly to light compound nuclei. The reaction products are
referred to as evaporation residues. These can be observed directly by counter tele-
scopes or mica track detectors. Very heavy compound nuclei with small fission
barriers (cf. fig. 3.5.2) decay preferentially by fission. The fission products can
be measured. For intermediate compound masses both the evaporation residues and the
fission fragments must be measured to obtain the cross-section σ_{CN} for compound-
nucleus formation. For heavy colliding nuclei it is sometimes difficult to distinguish the compound-nucleus fission from the deeply inelastic collision
(cf. chapter 2). Although the lifetime of the compound nucleus is large compared
to the duration of deeply inelastic collisions, the angular distribution of the
fragments does not always allow a clear distinction between the two components.
Apart from this experimental difficulty there are additional problems with the
correct interpretation of the experimental data on fusion and compound-nucleus
formation. We try to classify the experiments in the following way:

(i) Collisions between light nuclei with $E_{CM} > V_{CB}$.

(ii) Collisions at higher energies and/or between heavier nuclei where the effect
of a critical angular momentum ℓ_{crit}^{f} of the compound system is observed.

(iii) Collisions showing entrance channel effects.

(iv) Collisions where fusion and compound-nucleus formation are not found.

We discuss these different collisions in turn.

(i) All nuclear reactions between light HI lead practically completely to compound-
nucleus formation,

$$\sigma_{CN} \approx \sigma_{ER} \approx \sigma_{R} \tag{3.10.1}$$

if the energy is close to the Coulomb barrier V_{CB} . Since compound-nucleus fission

---

[†] This fission process has to be distinguished from the instability against prompt
fission discussed in section 3.5 . The latter prevents the formation of the
compound nucleus altogether.

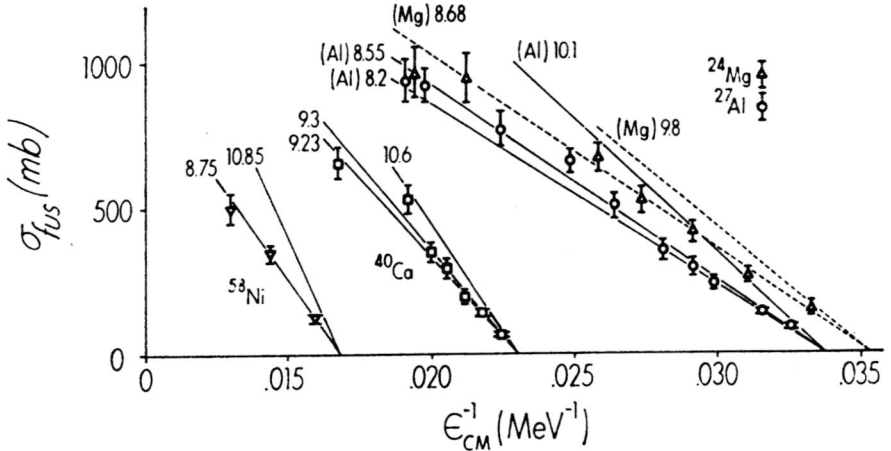

Fig. 3.10.1 Measured fusion cross section ($\sigma_{fus} \equiv \sigma_{ER}$) for the bombardment of ^{58}Ni, ^{40}Ca, ^{24}Mg and ^{27}Al by ^{32}S near the Coulomb barrier. (From [GU 73]).

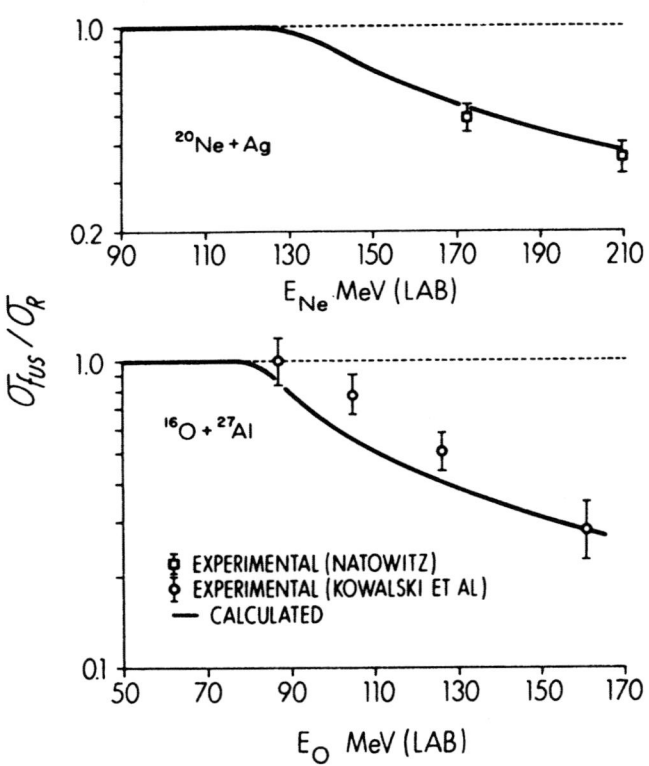

Fig. 3.10.2 Experimental and calculated fusion cross section ($\sigma_{fus} \equiv \sigma_{ER}$) for ^{20}Ne + Ag and ^{16}O + ^{27}Al. (From BL 72a).

plays no role for small energies and light compound nuclei, the cross-section σ_{ER}
for the evaporation residues represents the total compound-nucleus cross section
σ_{CN} .Fig. 3.10.1 shows some results which have been obtained in the bombardment of
^{58}Ni, ^{40}Ca, ^{27}Al and ^{24}Mg by ^{32}S . The cross-sections accurately determine R and
V_{CB} of eq. (3.3.1). The numbers attached to the straight lines give R in fm .

(ii)   In collisions at higher energies and/or for heavier nuclei where ℓ_{gr} becomes
close to or larger than ℓ^{f}_{crit} (cf. section 3.5), we have in general

$$\sigma_{fus} \equiv \sigma_{ER} < \sigma_{CN} < \sigma_{R} \quad . \tag{3.10.2}$$

Fig. 3.10.2 shows results for ^{20}Ne + Ag and ^{16}O + ^{27}Al . With increasing energy
σ_{ER}/σ_{R} decreases. This behaviour is well understood [BL 72a] in terms of the critical
angular momentum ℓ^{f}_{crit} , cf. section 3.6 and fig. 3.6.2 . The fusion cross-section
σ_{ER} is limited by prompt fission instability for $\ell > \ell^{f}_{crit}$ and by compound-
nucleus fission for $\ell < \ell^{f}_{crit}$, cf. fig. 3.6.1 . The significance of a critical

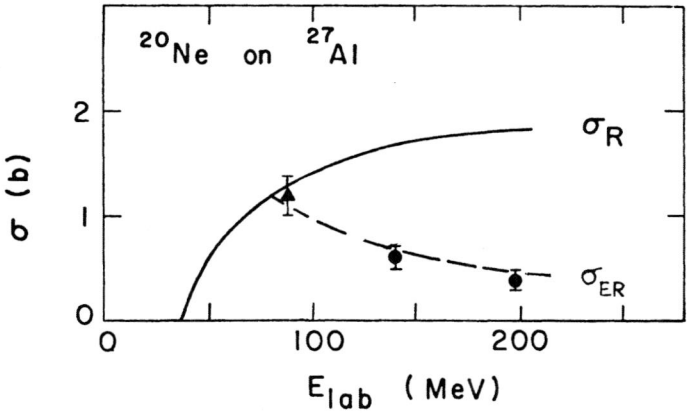

Fig. 3.10.3 Experimental σ_{ER} for ^{20}Ne + ^{27}Al . The solid curve
represents σ_{R} as given by eq. (3.3.1). The dashed
curve gives the expected limit due to fission competition (From [PÜ 72]).

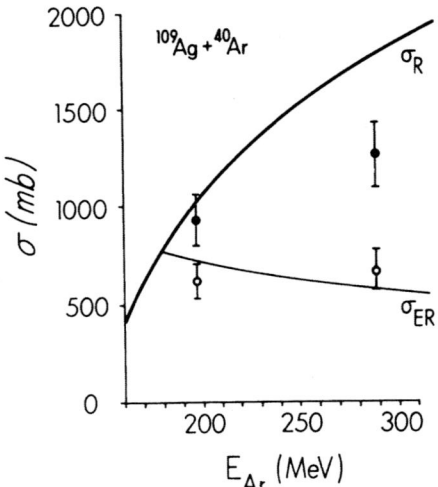

Fig. 3.10.4   The evaporation-residue cross section σ_{ER} (open circles)
and the compound-nucleus cross section σ_{CN} (black dots)
which includes the compound-nucleus fission decay. (From [PL 74b]).

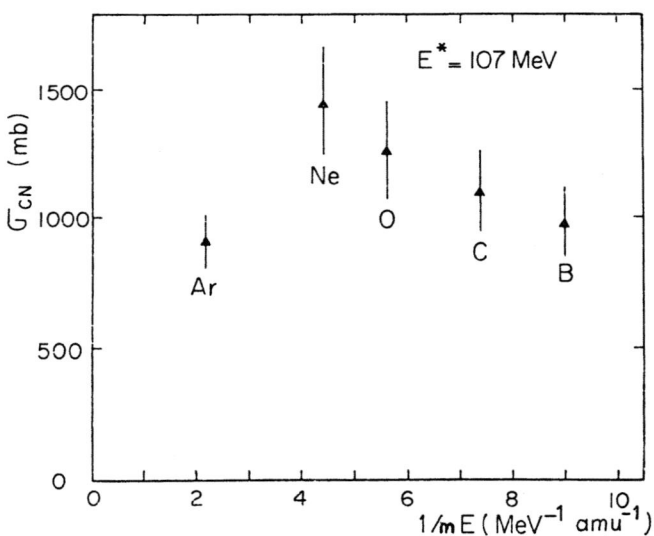

Fig. 3.10.5 Plot of σ_{CN} versus $(mE)^{-1}$ for rare-earth compound
nuclei at $E^{*} \approx 107$ MeV, (From [LE 75b]).

angular momentum for σ_{ER} is directly displayed in fig. 3.10.3 for $^{20}Ne + ^{27}Al$.
Fig. 3.10.4 illustrates the connection between σ_{ER}, σ_{CN} and σ_R for ^{40}Ar and
^{109}Ag . The thin line gives the calculated upper bound [PL 74b] for the cross-
section σ_{ER} of evaporation residues due to fission competition. The inclusion of
the compound-nucleus fission cross-section gives σ_{CN} . As expected we find
$\sigma_{CN} = \sigma_R$ for ℓ_{gr} considerably smaller than ℓ^f_{crit} . However, because of prompt
fission instability σ_{CN} becomes smaller than σ_R as soon as ℓ_{gr} becomes larger
than ℓ^f_{crit} .

(iii) Although the concept of a critical angular momentum of the compound system
has been successfully applied to various experimental data on σ_{CN} and σ_{ER} , it
turns out that it is not always applicable: There are experimental data which show
a dependence of the compound-nucleus formation on the dynamics of the HI collisions.
This is shown by entrance-channel effects on σ_{CN} for the formation of the same
compound nucleus at the same excitation energy with different projectiles and targets.
If σ_{CN} were limited by the same value of ℓ^f_{crit} in all cases, we would have
$\sigma_{CN} \propto \lambda^2 \propto (mE_{CM})^{-1}$ according to eq. (3.3.7). Here, m is the reduced mass in the

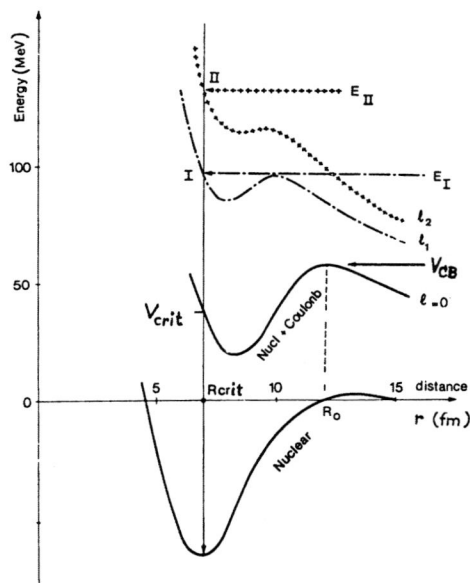

Fig. 3.10.6 Schematic diagram for the definition of R_{crit} and
V_{crit} . (From [LE 75b]).

entrance channel. The experimental points do not at all show this $(mE)^{-1}$ dependence.

To explain this dependence of σ_{CN} on the entrance channel, some authors assume

that compound-nucleus formation takes place for all angular momenta $\ell < \ell_{max}$.

Here, ℓ_{max} depends on the entrance channel and is related to the critical angular

momentum (differing from ℓ_{crit}^{f}) introduced in subsection 2.3.3 [BA 73, WI 73].

A different point of view has been taken by Lefort [LE 75a,b] who introduced a

critical distance R_{crit}, see fig. 3.10.6. It is assumed that the colliding nuclei

have to come closer than this distance in order that compound-nucleus formation takes

place. According to fig. 3.10.6 we define an ℓ value ℓ_1 for which the effective

potential V_{ℓ_1} taken at $r = R_{crit}$ becomes equal to the energy of the outer barrier.

For $\ell < \ell_1$ the effective potential $V_\ell(r)$ at $r = R_{crit}$ is always smaller than

the corresponding barrier energy. In order to reach R_{crit} the nuclei have only to

cross the barrier. Hence, for $E < V_{\ell_1}(R_{crit}) \equiv E_I$ (cf. fig. 3.10.6) the compound-

nucleus cross-section is determined by eq. (3.3.1) if we take a fixed position R

for the effective Coulomb barriers. The situation changes qualitatively for $E > E_I$.

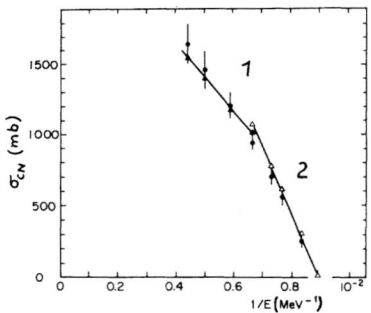

Fig. 3.10.7 Plot of σ_{CN} versus $1/E$ for $^{40}Ar + {}^{118}Sn$. The black dots are
the experimental values. Curve 1 is obtained from eq.(3.10.4)
with $R_{crit} = 1.05\ (A_1^{1/3} + A_2^{1/3})$ fm and $V_{crit} = 90$ MeV.
Curve 2 is obtained from eq. (3.3.1) with $R = 1.4\ (A_1^{1/3}+A_2^{1/3})$
and $V_{CB} = 112$ MeV. (From [LE 75a]).

The Coulomb barrier does no longer limit the compound-nucleus cross-section σ_{CN}.
Instead, the critical distance R_{crit} with the corresponding potential V_{crit} limits
σ_{CN} . Therefore, R_{crit} and V_{crit} replace the corresponding quantities R and V_{CB}
in eq. (3.3.1),

$$\sigma_{CN} = \pi R_{crit}^2 \left(1 - \frac{V_{crit}}{E_{CM}} \right) \tag{3.10.3}$$

for $E > E_I$. The value for ℓ_1 depends on the form of the combined nuclear and
Coulomb potentials. For very large ℓ_1 , E_I becomes very large. For $\ell_1 = 0$ one
can only observe a variation of σ_{CN} as given by eq. (3.10.3). For the collision
$^{40}Ar + {}^{118}Sn$ a transition as described above has been observed, cf. fig. 3.10.7 .
The energy E_I is approximately 150 MeV ($E_I^{-1} \approx 0.67 \cdot 10^{-2}$ MeV^{-1}) in this case.
A generalized expression for the cross-section of compound-nucleus formation has
been obtained by Glas and Mosel [GL74, 75] . This expression contains eqs. (3.3.1)
and (3.10.3) as limits for energies close to the Coulomb barrier and far above the
Coulomb barrier, respectively.

(iv) In collisions between very heavy nuclei, like Kr or Xe on Pb or U , neither
fusion nor compound-nucleus formation have been observed. The main part of the cross-
section is due to deeply inelastic collisions. The classical equation of motion with
friction forces is considered as a reasonable starting point for describing this
process (cf. chapter 2). More generally, relaxation phenomena observed in deeply
inelastic collisions have been interpreted within non-equilibrium statistical
mechanics (cf. section 8.4).

References

BA 73   R. Bass, Phys. Lett. $\underline{47B}$ (1973) 139

BE 61   R. Beringer and W.J. Knox, Phys. Rev. $\underline{121}$ (1961) 1195

BE 71   M.C. Bertin, S.L. Tabor, B.A. Watson, Y. Eisen and G. Goldring,
Nucl. Phys. $\underline{A167}$ (1971) 216

BE 73   F.D. Becchetti, P.R. Christensen, V.I. Manko and R.J. Nickles,
Nucl. Phys. $\underline{A203}$ (1973) 1

BJ 74   S. Björnholm, A. Bohr and B.R. Mottelson, in Physics and Chemistry of
Fission, 1973 (IAEA, Vienna,1974), vol. 1, p.367

BL 72a M. Blann and F. Plasil, Phys. Rev. Lett. $\underline{29}$ (1972) 303

BL 72b C. Bloch in ref. $\left[\text{GA 72}\right]$, p. 379

BO 39   N. Bohr and J.A. Wheeler, Phys. Rev. $\underline{56}$ (1939) 426

BO 69   A. Bohr and B.R. Mottelson, Nucl.Structure (Benjamin, New York, 1969)vol.I

BO 71   M. Böhning, Habilitationsschrift 1971, TU München, unpublished.

CH 71   F.S. Chang, J.B. French and T.H. Thio, Ann.Phys. (N.Y.) $\underline{66}$ (1971) 137

CO 63   S. Cohen, F. Plasil and W.D. Swiatecki, Proc. of the Third Conf. on
Reactions between Complex Nuclei, ed. by A. Ghiorso et al. (University of
California Press, Berkeley - Los Angeles, 1963) p.325

CO 74   S. Cohen, F. Plasil and W.J. Swiatecki, Ann.Phys. (N.Y.)$\underline{82}$ (1974) 557

FR 72   J.B. French and F.S. Chang, in ref. $\left[\text{GA 72}\right]$ p.405

GA 72   Statistical Properties of Nuclei, ed. by J.B. Garg (Plenum Press, New York -
London, 1972)

GL 74   D. Glas and U. Mosel, Phys. Rev. $\underline{C10}$ (1974) 2620

GL 75   D. Glas and U. Mosel, Nucl. Phys. $\underline{A237}$ (1975) 429

GR 67   J.R. Grover, Phys. Rev. $\underline{157}$ (1967) 832

GU 73   H.H. Gutbrod, W.G. Winn and M. Blann, Nucl.Phys. $\underline{A213}$ (1973) 267

HU 72   J.R. Huizenga, in ref. $\left[\text{GA 72}\right]$ p.425

JO 73   A. Johnson and Z. Szymanski, Phys.Rep. $\underline{C7}$ (1973) 181

KA 61   R. Kaufmann and R. Wolfgang, Phys.Rev. $\underline{121}$ (1961) 192

KA 66   I. Kanestrøm , Nucl. Phys. $\underline{83}$ (1966) 380

KA 68   I. Kanestrøm , Nucl. Phys. $\underline{A109}$ (1968) 625

LA 63   D.W. Lang, Proc. of the Third Conf. on Reactions between Complex Nuclei,
         Asilomar, 1963, ed. by A. Ghiorso et al. (University of California Press,
         Berkeley, 1963) p.248

LE 72   M. Lefort, J. de Physique $\underline{33}$ (1972) C5 - 73

LE 75a   M. Lefort, in Classical and Quantum Mechanical Aspects of Heavy Ion
         Collisions, ed. by H.L. Harney et al. (Springer-Verlag, Berlin-Heidelberg-
         New York, 1975) p.275

LE 75b   M. Lefort, preprint (1975) to be published in Heavy Ion Collisions,
         ed. by R.Bock

MY 66   W.D. Myers and W.J. Swiatecki, Nucl. Phys. $\underline{81}$ (1966) 1

NI 72   J.R. Nix, Los Alamos preprint, LA-DC-72-769 (1972)

PL 74a   F. Plasil, Proc. of the Int.Conf. on Reactions between Complex Nuclei,
         Nashville, 1973 (North-Holland Publ.Co., Amsterdam-Oxford, 1974) p.107

PL 74b   F. Plasil, preprint (Oak Ridge Nat. Lab., 1974) OR-NL-TM-4599

PÜ 72   F. Pühlhofer and R.M. Diamond, Nucl. Phys. $\underline{A191}$ (1972) 561

RO 57   N. Rosenzweig, Phys. Rev. $\underline{105}$ (1957) 950, ibid. $\underline{108}$ (1957) 817

SP 63   D. Sperber, Phys. Rev. $\underline{130}$ (1963) 468; ibid. $\underline{138}$ (1965) B 1028

SW 72   W.J. Swiatecki, J. de Physique $\underline{33}$ (1972) C5-45

TH 68   T.D. Thomas, Ann. Rev. Nucl. Science $\underline{18}$ (1968) 343

WI 63   B. Wilkins and G. Igo, Proc. of the Third Conf. on Reactions between
         Complex Nuclei, Asilomar, 1963, ed. by A. Ghiorso et al. (University of
         California Press, Berkeley, 1963) p.241

WI 64   L. Wilets, Theories of Nuclear Fission (Clarendon, Oxford, 1964)

WI 73   J. Wilczynski, Nucl. Phys. $\underline{A216}$ (1973) 386

4.   Some elements of nuclear scattering theory

It is the purpose of this chapter to lay the theoretical foundation for the developments to follow in chapters 5 to 8 . Although we collect here some of the formulas needed later, it is not our main aim to present a short introduction to formal scattering theory. On the contrary, we try to explain the physical concepts associated with the optical model, direct reactions, compound-nucleus processes etc. in a rather simple and intuitive fashion. This leads us to the decomposition of cross sections into direct and fluctuating parts which is rather fundamental for all that follows. For simplicity of presentation, we begin in section 4.1 with elastic scattering. The extension of the concepts to inelastic scattering in section 4.2 is then quite straightforward.

4.1   Elastic scattering

We recall the definition (3.1.1) of a channel. For the theoretician, elastic scattering really means that incident and outgoing channels are the same. Reactions in which only the channel spin s   and/or only the relative angular momentum $\ell = \ell_{12}$ change, without intrinsic excitation of either fragment, are not considered elastic in this sense. Practically nothing is known about such processes in HI scattering, and there is no reason at the moment to worry about these subtle differences. We rather neglect the spins of both projectile and target altogether and write the elastic scattering amplitude for spherically symmetric potentials in the form [ME 61, chapter X]

$$ f(\theta) = \frac{1}{2ik} \sum_{\ell=0}^{\infty} (2\ell+1)\,(S_\ell - 1)\, P_\ell (\cos \theta) \qquad (4.1.1) $$

where θ is the c.m. scattering angle, and $S_\ell = \exp(2i\phi_\ell)$ is the diagonal element of the S-matrix with the complex scattering phase shift ϕ_ℓ pertaining to angular momentum ℓ . The minus one in eq. (4.1.1) results from the subtraction of the plane wave part which exists even in the absence of any scatterer. The c.m. momentum is given by $\hbar k$. For scattering by a real spherical potential, conservation of probability implies $|S_\ell| = 1$ and, hence, ϕ_ℓ real. For scattering of composite particles where inelastic processes are possible, some flux is lost from the elastic channel.

Hence, $|S_\ell| < 1$. Writing $\phi_\ell = \Phi_\ell + i\, \Xi_\ell$ with Φ_ℓ, Ξ_ℓ real we thus find $\Xi_\ell > 0$.

The c.m. elastic cross section is given by

$$\frac{d\sigma_{el}}{d\Omega} = \left| f(\theta) \right|^2 . \tag{4.1.2}$$

The optical theorem $\left[\text{ME 62, eq. (XIX.187)}\right]$ allows us also to find the total cross section. It is given by

$$\sigma_{tot} = \frac{4\pi}{k}\, \mathfrak{Im}\, f(0) = \frac{2\pi}{k^2} \sum_{\ell=0}^{\infty} (2\ell+1)\,(1 - \mathfrak{Re}\, S_\ell) . \tag{4.1.3}$$

The integrated elastic cross section has the value

$$\sigma_{el} = \int d\Omega\, \left|f(\theta)\right|^2 = \frac{\pi}{k^2} \sum_{\ell=0}^{\infty} (2\ell+1)\, \left|S_\ell - 1\right|^2 . \tag{4.1.4}$$

The reaction cross section σ_R is the difference (cf. eq.(3.3.8))

$$\sigma_R = \sigma_{tot} - \sigma_{el} = \frac{\pi}{k^2} \sum_{\ell=0}^{\infty} (2\ell+1)\,(1 - |S_\ell|^2) . \tag{4.1.5}$$

We also recall that for the elastic scattering of identical particles, only the terms even or odd in ℓ appear, depending on the channel spin s . More precisely, we have

$$\frac{d\sigma_{el}}{d\Omega} = \left| f(\theta) + (-)^s f(\pi - \theta) \right|^2 \tag{4.1.6}$$

valid for Bosons and Fermions .

We now turn to the energy dependence of $S_\ell(E)$ or equivalently, of $\phi_\ell(E)$. It has mainly two causes: (i) There is a threshold behaviour determined by Coulomb penetration factors. This is well-understood and of no concern to us at present. (ii) In the domain of c.m. energies E and angular momentum values ℓ where compound-nucleus formation is possible, ϕ_ℓ is a rapidly changing function of energy.

To demonstrate point (ii), we first take the case of an isolated resonance. The S-matrix element has the familiar Breit-Wigner form

$$\exp\left(2i\phi_\ell\right) = \exp\left(2i\phi_\ell^o\right)\left(1 - i\ \frac{\Gamma_{e\ell}}{E - E_o + \tfrac{1}{2}i\Gamma_{tot}}\right) \qquad (4.1.7)$$

with the background phase-shift ϕ_ℓ^o which is slowly varying with energy. The quantities E_o, Γ_{tot} and Γ_{el} denote, respectively, the resonance energy, the total decay width and the partial width for the decay of the compound state into the elastic channel. The dependence of $\exp\left(2i\phi_\ell\right)$ upon energy is displayed in the "Argand" diagram of fig. 4.1.1 where the trajectory of $\exp\left(2i\phi_\ell\right)$ as a function of E is plotted in the complex plane. According to eq. (4.1.7), ϕ_ℓ increases by π over the resonance, and, therefore, $\exp\left(2i\phi_\ell\right)$ describes a circle in a counterclockwise direction around the origin. This circle coincides with the unit circle if the resonance is "elastic", i.e. if $\Gamma_{el} = \Gamma_{tot}$. As the ratio Γ_{el}/Γ_{tot} decreases, the radius of the circle shrinks.

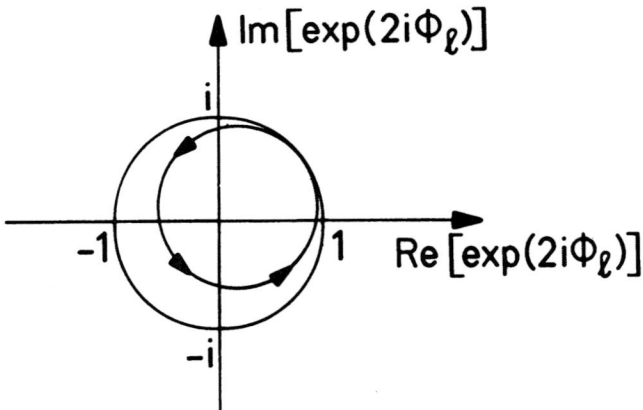

Fig. 4.1.1 "Argand" diagram in the resonance energy region.

In the domain of c.m. energies where compound-nucleus formation is possible, the density of the compound states is usually very high unless ℓ is so large that we are close to the Yrast line (cf. section 3.II). Except for this case, we deal with a large number of resonances of the form (4.1.7) per unit energy interval, each compound state corresponding to a resonance, and it is quite impossible to account for the detailed features of all these resonances. In this situation, it makes no sense to consider in detail the energy dependence of $\phi_\ell(E)$, and one focusses attention on the <u>average</u> <u>over</u> <u>energy</u> of $\exp(2i\phi_\ell)$, denoted by $<\exp(2i\phi_\ell)>$. The average is taken over an energy interval containing a large number $N \gg 1$ of resonances, and it is hoped that $<\exp(2i\phi_\ell)>$ is a smooth and simple function of energy although $\phi_\ell(E)$ is not. It is the aim of the <u>optical model</u> of elastic scattering to account for the behaviour of the quantities $<\exp(2i\phi_\ell)> = <S_\ell>$. More precisely, the optical model potential $V_{opt}(r)$ is defined by the requirement that the solution $\psi_\ell(r)$ of the radial Schrödinger equation

$$\left\{-\frac{\hbar^2}{2m}\frac{d^2}{dr^2} + \frac{\hbar^2 \ell(\ell+1)}{2mr^2} + V_{Clb}(r) + V_{opt}(r) - E\right\} \psi_\ell(r) = 0 \qquad (4.1.8)$$

has the asymptotic behaviour given by the phase shift ϕ_ℓ^{opt} such that

$$\exp(2i\phi_\ell^{opt}) \equiv S_\ell^{opt} = \langle S_\ell \rangle \equiv \langle \exp(2i\phi_\ell) \rangle. \qquad (4.1.9)$$

Looking at the Argand diagram of fig. 4.1.1 we see that averaging S_ℓ over a single resonance, we get a mean value located somewhere within the unit circle. The same holds true a fortiori for the average over <u>many</u> resonances. This shows that $|<\exp(2i\phi_\ell)>|$ will, in general, be located well within the unit circle. In particular, we have $|<\exp(2i\phi_\ell)>| < 1$ even if $|\exp(2i\phi_\ell)| = 1$ for all E in the averaging interval. This shows that in the optical model, absorption is due to two physically unrelated causes: (i) ϕ_ℓ itself is complex with $\mathcal{I}m\,\phi_\ell = \Xi_\ell > 0$ because other channels besides the elastic one are open; (ii) the price we pay for obtaining a smooth energy dependence of $<S_\ell>$ is a reduction of the magnitude of S_ℓ by the averaging process. This is also described as "absorption" in the context of the optical model. Explicit formulas for the optical model may be found in a more general context in subsection 7.3.2 .

The elastic cross section (4.1.2) changes as rapidly with energy as the $S_\ell(E)$ do, and again it makes sense to consider the average over energy of $d\sigma_{el}/d\Omega$. In order to do so, we write

$$S_\ell(E) = \langle S_\ell(E) \rangle + S_\ell^{f\ell}(E) , \tag{4.1.10a}$$

and, correspondingly,

$$f(\theta) = \langle f(\theta) \rangle + f^{f\ell}(\theta) . \tag{4.1.10b}$$

The superscript fl indicates the "fluctuating part" which varies rapidly and by definition has mean value zero. Inserting eq. (4.1.10b) into eq. (4.1.2) we find

$$\begin{aligned}
\langle \frac{d\sigma_{el}}{d\Omega} \rangle &= |\langle f(\theta) \rangle|^2 + \langle |f^{f\ell}(\theta)|^2 \rangle \\
&= \frac{d\sigma_{el}^{opt}}{d\Omega} + \langle |f^{f\ell}(\theta)|^2 \rangle \equiv \frac{d\sigma_{el}^{opt}}{d\Omega} + \langle \frac{d\sigma_{el}^{f\ell}}{d\Omega} \rangle
\end{aligned} \tag{4.1.11}$$

where we have used eq. (4.1.9). For the average total cross section, we have because of eqs. (4.1.3) and (4.1.9)

$$\langle \sigma_{tot} \rangle = \sigma_{tot}^{opt} . \tag{4.1.12}$$

Finally, the reaction cross sections obey the relationship

$$\begin{aligned}
\langle \sigma_R \rangle &= \langle \sigma_{tot} \rangle - \langle \sigma_{el} \rangle = \sigma_{tot}^{opt} - \sigma_{el}^{opt} - \langle \sigma_{el}^{f\ell} \rangle \\
&= \sigma_R^{opt} - \langle \sigma_{el}^{f\ell} \rangle ,
\end{aligned} \tag{4.1.13}$$

or

$$\sigma_R^{opt} = \langle \sigma_R \rangle + \langle \sigma_{el}^{fl} \rangle \quad .$$ (4.1.14)

Eq. (4.1.14) shows that the mean value of the elastic fluctuating cross section is considered as part of the <u>reaction</u> <u>cross-section</u> in the context of the optical model. This explains the origin of the "additional absorption" mentioned under (ii) above. The optical model potential does not aim at reproducing the full elastic scattering cross-section, but only part of it. σ_R^{opt} agrees with $\langle \sigma_R \rangle$, and $d\sigma_{el}^{opt}/d\Omega$ with $\langle d\sigma_{el}/d\Omega \rangle$, only if the mean fluctuation part of $d\sigma_{el}/d\Omega$ is negligible.

Friedman and Weisskopf [FR 55] have given an illustrative interpretation of the two contributions $d\sigma_{el}^{opt}/d\Omega$ and $\langle d\sigma_{el}^{fl}/d\Omega \rangle$ to the average of the elastic cross section. Taking the average of $\Psi_\ell(r)$ over an energy interval I which contains very many resonances and, therefore, is large compared to the mean distance and to the mean total width $\langle \Gamma_{tot} \rangle$ of these resonances, corresponds to considering a wave packet of (time length) $\Delta t = 2\pi\hbar/\Delta E$ impinging on the target. Part of the wave packet is scattered immediately, i.e. with a time delay smaller than or comparable with Δt. It is this fast part which is considered in the optical model. The other part of the wave packet populates the compound nucleus which decays in a characteristic time $2\pi\hbar/\langle \Gamma_{tot} \rangle \gg \Delta t$. Such slow time dependences correspond to rapid variations in the Fourier- transformed representation, i.e. the energy representation, and, therefore, do not contribute to the mean value of $\langle f(\theta) \rangle$ which alone is considered in the optical model. This simple consideration gives the clue to several features of the optical model. First, S_ℓ^{opt} accounts only for the <u>fast</u> part of elastic scattering. The fast part of the reaction should be dominates by <u>few</u> degrees of freedom. Hence the simplicity of the usual parametrization of the optical model. The simplest such degree of freedom is just the shape of the target nucleus. Therefore, $d\sigma_{el}^{opt}/d\Omega$ is often referred to as "shape-elastic" scattering. Second, the contribution $\langle d\sigma_{el}^{fl}/d\Omega \rangle$ is, in the light of this interpretation, due to a slow process, i.e. due to the decay of the compound nucleus back into the elastic channel. For this reason, it is often referred to as "compound-elastic" scattering.

The introduction of the optical model is obviously useful because it simplifies the description of the reaction. However, it leaves us with the problem of finding the compound-elastic scattering cross-section. At high excitation energies, where very many channels are open, the compound nucleus populates, aside from barrier penetration effects, all these channels uniformly, and decay back into the elastic channel is negligible. This is the rule for HI reactions. For nucleon-induced reactions, the situation sometimes requires the evaluation of $<d\sigma_{el}^{fl}/d\Omega>$. This can be done with the help of the statistical theory of nuclear reactions, see chapter 8 .

In chapter 5 , we present a number of simple models which aim at a calculation of the mean value of the elastic scattering amplitude $<f(\theta)>$. All these models involve simple pictures, and few degrees of freedom, as we expect. They may all be viewed as approximations to a full optical-model description of elastic HI scattering.

4.2   Inelastic processes

The generalization of the considerations carried through in section 4.1 for elastic scattering to the general case is now quite straightforward. We first express the differential cross section in terms of the S-matrix elements. The latter carry the indices $(\alpha s \ell J)$ introduced in section 3.1 to denote a channel, where $\alpha = (\alpha_1, \alpha_2, J_1, J_2)$ defines the type of fragmentation and the intrinsic state of either fragment, s is the channel spin, and ℓ the relative orbital angular momentum. The S-matrix elements are written in the form $S^J_{\alpha's'\ell', \alpha s \ell}$ where the primed indices refer to the exit, the unprimed ones to the entrance channel. Following Blatt and Biedenharn [BL 52] we write the c.m. differential cross section for the reaction leading from state α to the state α' in the form

$$\frac{d\sigma_{\alpha\alpha'}}{d\Omega} = \sum_{ss'} \frac{\lambda_\alpha^2}{(2I+1)(2i+1)} \sum_{L=0}^{\infty} P_L(\cos\theta) \frac{(-)^{s'-s}}{4} \cdot$$

$$\cdot \sum_{\ell_1 \ell_2 \ell'_1 \ell'_2 J_1 J_2} Z(\ell_1 J_1 \ell_2 J_2; s L) \ Z(\ell'_1 J_1 \ell'_2 J_2; s'L) \cdot$$

$$\cdot Re\left\{ (\delta_{\alpha\alpha'} \delta_{\ell_1 \ell'_1} \delta_{ss'} - S^{J_1}_{\alpha's'\ell'_1, \alpha s \ell_1})(\delta_{\alpha\alpha'} \delta_{\ell_2 \ell'_2} \delta_{ss'} - S^{J_2 *}_{\alpha's'\ell'_2; \alpha s \ell_2}) \right\}. \qquad (4.2.1)$$

Here, I and i are the spins of projectile and target, respectively, θ is the c.m. scattering angle, λ_α is $\hbar$ times the inverse relative momentum in the entrance channel, and the Z's are geometrical coefficients given by

$$Z(\ell_1 \mathcal{J}_1 \ell_2 \mathcal{J}_2, sL) \equiv i^{L-\ell_1+\ell_2} \sqrt{(2\ell_1+1)(2\ell_2+1)(2\mathcal{J}_1+1)(2\mathcal{J}_2+1)} \cdot$$
$$\cdot W(\ell_1 \mathcal{J}_1 \ell_2 \mathcal{J}_2, sL)(\ell_1 \ell_2 00|L0). \tag{4.2.2}$$

As mentioned in section 4.1 , the expression (4.2.1) is a rapidly varying function of excitation energy in the domain where many compound resonances contribute, and it is useful to consider averages over energy over an interval containing a large number of such resonances. Denoting these averages by a bracket, we write

$$S_{ab}^{\mathcal{J}} = \langle S_{ab}^{\mathcal{J}} \rangle + S_{ab}^{\mathcal{J}\mathcal{l}}, \quad \langle S_{ab}^{\mathcal{J}\mathcal{l}} \rangle = 0 \tag{4.2.3}$$

in complete analogy to eqs. (4.1.10). The indices a and b are channel labels. The averaging interval in eq. (4.2.3) is denoted by I and must be chosen in such a way that

$$I \gg D, \langle \Gamma_{tot} \rangle \tag{4.2.4}$$

where D is the mean spacing and $\langle \Gamma_{tot} \rangle$ the mean total width of the resonances of given spin J and parity π . The quantity $\langle S_{ab}^J \rangle$ accounts for the fast part of the reaction just as $\langle S_\ell \rangle$ does in the elastic case, see section 4.1 . We, therefore, expect $\langle S_{ab}^J \rangle$ to invoke only a few degrees of freedom of the system, so that simple models can be used to calculate it. While $\langle S_{aa}^J \rangle$ is given by the optical model, $\langle S_{ab}^J \rangle$ with a $\neq$ b can be calculated from the various models for <u>direct reactions</u> described in chapter 7. Because of the strong absorption, such direct reactions mostly take place at the grazing distance, and this is in nice agreement with fig. 3.4.4 .

Averaging the expression (4.2.1) over energy and using eq. (4.2.3), we obtain

$$\left\langle \frac{d\sigma_{\alpha\alpha'}}{d\Omega} \right\rangle = \frac{d\sigma_{\alpha\alpha'}^{Direct}}{d\Omega} + \left\langle \frac{d\sigma_{\alpha\alpha'}^{fl}}{d\Omega} \right\rangle , \tag{4.2.5}$$

where $d\sigma_{\alpha\alpha'}^{Direct}/d\Omega$ has the form of eq. (4.2.1) with S_{ab}^{J} replaced by $\langle S_{ab}^{J} \rangle$. It corresponds to that part of the mean cross section which can be attributed to a direct reaction. The other part has the form

$$\left\langle \frac{d\sigma_{\alpha\alpha'}^{fl}}{d\Omega} \right\rangle = \sum_{ss'L} \frac{\lambda_{\alpha}^{2}}{(2I+1)(2i+1)} \frac{(-)^{s'-s}}{4} P_{L}(\cos\theta)$$

$$\cdot \sum_{\ell_{1}\ell_{2}\ell_{1}'\ell_{2}'\mathcal{J}_{1}\mathcal{J}_{2}} Z(\ell_{1}\mathcal{J}_{1},\ell_{2}\mathcal{J}_{2};sL)\, Z(\ell_{1}'\mathcal{J}_{1},\ell_{2}'\mathcal{J}_{2};s'L)$$

$$\cdot \left\langle Re \left\{ S_{\alpha's'\ell_{1}';\alpha s\ell_{1}}^{\mathcal{J}_{1},fl}\, S_{\alpha's'\ell_{2}';\alpha s\ell_{2}}^{\mathcal{J}_{2}fl}{}^{*} \right\} \right\rangle . \tag{4.2.6}$$

Following the interpretation described in section 4.1, we see that $\langle d\sigma_{\alpha\alpha'}^{fl}/d\Omega \rangle$ describes the compound-nucleus contribution to the mean cross section. In contra-distinction to the case of elastic scattering, where the shape elastic contribution almost always dominates, $\langle d\sigma_{\alpha\alpha'}^{fl}/d\Omega \rangle$ is by no means negligible in comparison with $d\sigma_{\alpha\alpha'}^{Direct}/d\Omega$. This is the case particularly when entrance and exit channel contain very different fragmentations and/or configurations, so that α' cannot be reached from α through a collision involving only few degrees of freedom. It is, therefore, important to calculate $\langle \sigma_{\alpha\alpha'}^{fl}/d\Omega \rangle$, and means to do this are provided by the statistical model detailed in chapter 8 .

References

BL 52   J.M. Blatt and L.C. Biedenharn, Revs.Mod.Phys. 24 (1952) 258

FR 55   F.L. Friedman and V.F. Weisskopf, in "Niels Bohr and the Development of Physics", ed. W. Pauli (Pergamon Press, New York 1955) p.134

ME 61   A. Messiah, Quantum Mechanics, vol. 1 (North-Holland Publishing Co., Amsterdam, 1961)

ME 62   A. Messiah, Quantum Mechanics, vol. 2 (North-Holland Publishing Co., Amsterdam, 1962)

5.   Elastic scattering

In most HI collisions the local wave length $\lambdabar(r)$ of relative motion is small as compared to the nuclear interaction region R . Therefore, classical approximations have become extremely useful in describing and understanding HI reactions. It is for this reason that the classical theory has been described separately in chapter 2 . But according to section 2.4 , the classical treatment is limited by quantal effects (interferences and $|\text{grad } \lambdabar(r)| \approx 1$) and by absorption. The quantal effects on the classical scattering are described in the semi-classical approximation (section 5.1). The optical model (cf. section 4.1) is applied to the elastic HI scattering in section 5.2 . The following three sections are concerned with various parametrizations of the scattering phase shifts (so-called parametrized phase-shift models). These may be viewed as simplifications of or approximations to the optical model. In the sharp cut-off diffraction model (section 5.3) the importance of strong absorption is stressed, leading to Fresnel and Fraunhofer diffraction patterns. This description is generalized in the smooth cut-off model (section 5.4). The Regge-pole model (section 5.5) can account for specific structures in HI scattering which are related to surface waves or potential resonances.

5.1   Semiclassical theory of elastic scattering $\left[\text{FO 59, MC 70, BR 72b}\right]$

As we shall see, the quantum-mechanical scattering phase shifts can be simply related to the classical deflection function when the conditions for a semi-classical description of the quantum-mechanical scattering are met. The main difference to the classical treatment is due to interference effects from different orbits. Instead of summing up the cross sections as done in eq. (2.1.12) one has to sum the amplitudes. Therefore, the classical rainbow and glory scattering are modified significantly, but the behaviour of the cross section can still be interpreted with the help of the classical deflection function.

The following treatment is restricted to real potentials only. Since absorption is strong in HI collisions, one has to be careful in applying the semi-classical results to experimental data. Absorption may be taken into account to some extent as described in subsections 2.4.2 and 5.1.3 . But there are additional effects (diffraction, cf. sections 5.3 and 5.4) which may even dominate the elastic cross section.

5.1.1 Derivation of the semiclassical scattering amplitude

The approximations which lead from the quantum-mechanical scattering amplitude given by eq. (4.1.1) to the semiclassical approximation can be divided into
four successive steps:

(i)   Evaluate the scattering phase-shifts ϕ_ℓ in the JWKB method.

(ii) Use the asymptotic expression of the Legendre polynomials for $\ell \gg 1$.

(iii) Replace the sum over ℓ in the scattering amplitude by an integral over ℓ .

(iv) Evaluate the expression by the method of stationary phase.

(i)   JWKB method for the calculation of scattering phase-shifts (cf. [ME 61, ch.VI]
      and refs. given above)

We consider a spherically symmetric potential $V(r)$. The radial wave equation
reads

$$y_\ell'' + k_\ell^2(r)\, y_\ell(r) = 0 \qquad (5.1.1)$$

where

$$k_\ell(r) = \sqrt{\frac{2m}{\hbar^2}\left[E - V(r) - \frac{\hbar^2 \ell(\ell+1)}{2mr^2}\right]} \qquad (5.1.2)$$

is the local radial wave number for the partial wave of angular momentum ℓ . Inserting

$$y_\ell(r) = A_\ell(r)\, exp\left[i\, S_\ell(r)/\hbar \right] \qquad (5.1.3)$$

into (5.1.1) with $A_\ell(r)$ and $S_\ell(r)$ real, we get the coupled differential equations

$$\left(\frac{S_\ell'}{\hbar}\right)^2 - k_\ell^2 = \frac{A_\ell''}{A_\ell} \qquad (5.1.4)$$

$$\frac{A_\ell'}{A_\ell} = -\frac{1}{2}\frac{S_\ell''}{S_\ell'} \qquad (5.1.5)$$

for the real and imaginary parts. The integration of eq. (5.1.5) gives

$$A_\ell(r) = const\ \frac{1}{\sqrt{|S_\ell'|}} \quad . \qquad (5.1.6)$$

In eq. (5.1.4) we neglect A_ℓ''/A_ℓ which amounts to the neglect of the terms S_ℓ'''/S_ℓ'
and $(S''/S')^2$. In this approximation we get

$$S_\ell'(r) = \pm\ \hbar k_\ell(r) \qquad (5.1.7)$$

where $k_\ell(r)$ denotes the radial wave number (5.1.2). We emphasize that eq. (5.1.7) is not the result of a consistent treatment of eq. (5.1.4) in the lowest order of $\hbar^2$, because the angular momentum barrier is treated here as though it were of zeroth order in $\hbar^2$. In contradistinction, the semiclassical treatment of the Schrödinger equation in three dimensions (cf. subsection 2.4.1) can be considered as a rigorous approximation to $S(r)$ in the lowest order of $\hbar^2$. Therefore, the three-dimensional treatment is substantially different from the present JWKB method for the radial wave equation. Inserting the JWKB solution

$$y_\ell(r) = \sqrt{\frac{1}{k_\ell(r)}} \; exp \left[\pm i \int_{a_\ell}^{r} k_\ell(r') \, dr' \right] \tag{5.1.8}$$

into eq. (5.1.1) we find the condition

$$\left| \frac{d}{dr} \lambda_\ell(r) \right| = \frac{\left| \frac{d}{dr} k_\ell(r) \right|}{\left| k_\ell^2(r) \right|} \ll 1 \quad \text{and} \quad \frac{\left| \frac{d^2}{dr^2} k_\ell(r) \right|}{\left| k_\ell^3(r) \right|} \ll 1 \tag{5.1.9}$$

for the validity of the JWKB method. These conditions are clearly violated near a classical turning point $r = a_\ell$ where $k_\ell(r = a_\ell)$ vanishes. In order to get the solution in the vicinity of $r = a_\ell$, one has to go back to the Schrödinger equation. Let $k_\ell^2(r) < 0$ for $r < a_\ell$ and $k_\ell^2(r) > 0$ for $r > a_\ell$. From the theory of singular differential equations (cf. [SC 55, ME 61, FR 65]) one finds the following connection formulae

$$y_\ell^{(1)}(r) = \begin{cases} \sqrt{\frac{1}{\lambda_\ell(r)}} \; exp \left[\int_{r}^{a_\ell} \lambda_\ell(r) \, dr' \right] & \text{for } r < a_\ell \\[3mm] -\sqrt{\frac{1}{k_\ell(r)}} \; sin \left[\int_{a_\ell}^{r} k_\ell(r') \, dr' - \frac{\pi}{4} \right] & \text{for } r > a_\ell \end{cases} \tag{5.1.10}$$

$$y_\ell^{(2)}(r) = \begin{cases} \frac{1}{2} \sqrt{\frac{1}{\lambda_\ell(r)}} \; exp \left[-\int_{r}^{a_\ell} \lambda_\ell(r') \, dr' \right] & \text{for } r < a_\ell \\[3mm] \sqrt{\frac{1}{k_\ell(r)}} \; cos \left[\int_{a_\ell}^{r} k_\ell(r') \, dr' - \frac{\pi}{4} \right] & \text{for } r > a_\ell \end{cases} \tag{5.1.11}$$

for the linearly independent functions $y_\ell^{(1)}$ and $y_\ell^{(2)}$. Here $\lambda_\ell = -ik_\ell$ has been used.

If only one turning point is present, i.e. if $k_\ell^2(r) < 0$ for all $r < a_\ell$, the JWKB solution to eq. (5.1.1) is given by $y_\ell^{(2)}$ alone, because $y_\ell^{(1)}$ is exponentially divergent for $r \to 0$. The corresponding phase shifts ϕ_ℓ are determined from

$$y_{\ell}^{(2)}(r) \overset{r \to \infty}{\sim} \frac{1}{k} \sin \left[(r - a_{\ell})k + \int_{a_{\ell}}^{r} dr' \{ k_{\ell}(r') - k \} + \frac{\pi}{4} \right] \tag{5.1.12}$$

where $k \equiv k_{\ell}(\infty)$ has been introduced. Hence,

$$\phi_{\ell} = \frac{\pi}{2} \left(\ell + \frac{1}{2} \right) - ka_{\ell} + \int_{a_{\ell}}^{\infty} \left[k_{\ell}(r') - k \right] dr' \tag{5.1.13}$$

are the JWKB phase shifts. The JWKB solution (5.1.11) fails to give the correct form for the wave function near the origin (angular momentum barrier!) even in the case of regular potentials. It has been observed by Langer that this is easily corrected by using an effective wave number

$$k_{\ell}^{eff}(r) \equiv \sqrt{ \frac{2m}{\hbar^2} \left[E - V(r) - \frac{\hbar^2 (\ell + \frac{1}{2})^2}{2mr^2} \right] } \tag{5.1.14}$$

instead of $k_{\ell}(r)$ where $\ell(\ell + 1)$ in eq. (5.1.2) is replaced by $(\ell + 1/2)^2$. With this Langer modification the JWKB phase shifts read

$$\phi_{\ell} = \frac{\pi}{2} \left(\ell + \frac{1}{2} \right) - ka_{\ell} + k \int_{a_{\ell}}^{\infty} dr' \left\{ \sqrt{ 1 - \frac{V(r')}{E} - \frac{\hbar^2 (\ell + \frac{1}{2})^2}{2mEr'^2} } - 1 \right\} \tag{5.1.15}$$

which is a good approximation even for small angular momenta ℓ if only the conditions (5.1.9) are satisfied for $r \neq a_{\ell}$.

The classical deflection function, eq. (2.1.4), and the semiclassical phase shifts, eq. (5.1.15) are closely related. To show this, we consider $\phi_{\ell}(k)$ as function of ℓ for fixed k. Taking the derivative of eq. (5.1.15) with respect to ℓ and comparing with eq. (2.1.4) we find

$$\Theta(L) = 2 \frac{\partial \phi_{\ell}}{\partial \ell} \tag{5.1.16}$$

if the classical angular momentum L is identified with $\hbar(\ell + \frac{1}{2})$. Thus the deflection function is determined by the derivative of the scattering phase shift with respect to ℓ. We emphasize that the relation (5.1.16) is valid but in cases where only one classical turning point occurs. Moreover, it should be realized that the relation (5.1.16) relates only to the asymptotic motion.

(ii)   In order to obtain a closed form for the semiclassical scattering amplitude
we consider the quantum-mechanical expression (4.1.1) for $\ell \gg 1$. The Legendre
polynomials are replaced by the asymptotic expressions valid for large ℓ ; in particular for $\sin \theta \gtrsim 1/\ell$ by

$$P_\ell(\cos \theta) \approx \frac{\sin\left[\left(\ell+\tfrac{1}{2}\right)\theta + \tfrac{\pi}{4}\right]}{\sqrt{\tfrac{\pi}{2}\left(\ell+\tfrac{1}{2}\right)\sin \theta}} \tag{5.1.17}$$

and for $\sin \theta \lesssim 1/\ell$ by

$$P_\ell(\cos \theta) \approx (\cos \theta)^\ell \, J_0\left(\left[\ell+\tfrac{1}{2}\right]\sin \theta\right) . \tag{5.1.18}$$

The ranges of validity of these formulas overlap slightly, so that the whole range of
θ is covered. The approximation formulas are fairly good even for rather small ℓ .

(iii) We now replace the sum over ℓ by an integral, i.e. $\sum_\ell \to \int_{\frac{1}{2}}^{\infty} d\lambda$ with
$\ell + \tfrac{1}{2} \to \lambda$. This is possible only if ϕ_ℓ is a smooth function of ℓ and
if many ℓ's contribute. For $\sin \theta \gtrsim 1/\ell$ this gives

$$f(\theta) \approx f_{sc}(\theta) = -\frac{1}{k\sqrt{2\pi \sin \theta}} \int_{\frac{1}{2}}^{\infty} d\lambda \sqrt{\lambda} \left\{ \exp\left[i\chi_+(\lambda)\right] - \exp\left[i\chi_-(\lambda)\right] \right\} \tag{5.1.19}$$

where we have used $\sum_\ell (2\ell+1) P_\ell(\cos\theta) = 0$ for $\theta \neq 0$ (contribution from incident
plane wave) and

$$\chi_\pm(\lambda) \equiv 2\phi(\lambda) \pm \lambda\theta \pm \tfrac{\pi}{4} . \tag{5.1.20}$$

(iv)   The integral (5.1.19) can be evaluated by using the method of stationary
phases. We can get a contribution either from the first exponential, i.e. for
$d\chi_+(\lambda)/d\lambda = 0$ which gives $d\phi(\lambda)/d\lambda = -\theta/2 - n\pi$, or from the second exponential,
i.e. for $d\chi_-(\lambda)/d\lambda = 0$ which gives $d\phi(\lambda)/d\lambda = +\theta/2 - n\pi$. Here $n\pi$ (with n integer)
takes care of the arbitrariness in the definition of the phases $\phi(\lambda)$, i.e., $\phi(\lambda)$

and $\phi(\lambda) + (\lambda - \frac{1}{2}) n \pi$ are equivalent (cf. also eq. (2.1.6)). The contribution to the scattering amplitude (5.1.19) from a stationary point λ_j is found to be

$$f_{sc}^{j}(\theta) = \sqrt{\frac{d\sigma_{c\ell}^{j}}{d\Omega}} \ exp \ (i\beta_j) \tag{5.1.21}$$

where $d\sigma_{c\ell}^{j} / d\Omega$ is given by eq. (2.1.10). The phase of the semiclassical scattering amplitude is determined by

$$\beta_j = 2\phi(\lambda_j) - \frac{\pi}{4} \ (2 - sgn \ \phi''(\lambda_j) - sgn \left[\phi'(\lambda_j) + n\pi\right] - 2\lambda_j \ \phi'(\lambda_j) . \tag{5.1.22}$$

If more than one trajectory contributes to a given angle, the total scattering amplitude is the sum of all contributions. The resulting interference patterns are strongly dependent on the phases β_j .

The semiclassical approximation uses the method of stationary phase. This is a good approximation, if for all essentially contributing λ values, $\phi(\lambda)$ can be approximated by a parabola near the stationary point λ_j . As is illustrated in fig. 5.1.1, the important region is where the tangent to $\phi(\lambda)$ differs from $\phi(\lambda)$ by less than a few radian. We also must have $\lambda_j \gg 1$ and $\Delta\lambda \gg 1$ where $\Delta\lambda$ is the width of the important region.

For purely repulsive potentials which are singular at the origin, like for instance the Coulomb potential, the deflection function $\Theta(\lambda) = 2 \ d\phi(\lambda)/d\lambda$ is monotonically decreasing from π to 0 with increasing λ (cf. fig. 2.1.2). Therefore, only one trajectory contributes to a specific scattering angle Θ and hence

$$\frac{d\sigma_{sc}}{d\Omega} = \left| f_{sc}(\theta) \right|^2 = \frac{d\sigma_{c\ell}}{d\Omega} \tag{5.1.23}$$

i.e. the semiclassical cross-section is equal to the classical cross-section. This is also true for an attractive potential, if $\Theta(\lambda)$ varies monotonically from $-\pi$ to 0 with increasing λ . The attractive Coulomb potential is the only simple attractive potential with this property. In general, there are several contributions

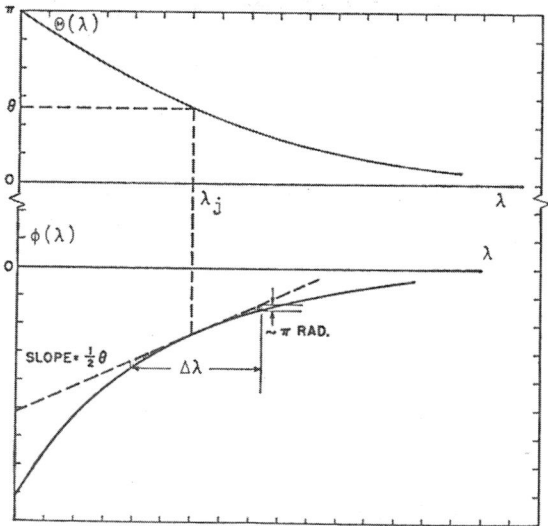

Fig. 5.1.1 Classical deflection function $\Theta(\lambda)$ and phase shift $\phi(\lambda)$
for illustrating the validity of the semiclassical
approximation. (From [FO 59]).

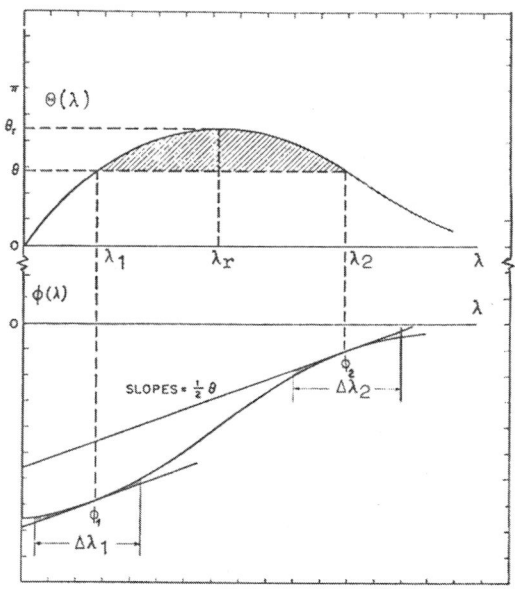

Fig. 5.1.2 Classical deflection function $\Theta(\lambda)$ and phase shift $\phi(\lambda)$
illustrating interference between two branches of
$\Theta(\lambda)$. The deflection function also shows a rainbow angle Θ_r at angular momentum λ_r . (From [FO 59]).

(5.1.21) to the semiclassical scattering amplitude from different trajectories which all satisfy the condition $\Theta + 2n\pi = \pm \theta$ with n integer. Therefore, the semi-classical cross-section

$$\frac{d\sigma_{sc}}{d\Omega} = \left| \sum_j \sqrt{\frac{d\sigma_{c\ell}^j}{d\Omega}} \; exp\,(i\beta_j) \right|^2 \tag{5.1.24}$$

takes into account the interference from the different trajectories. Consider the case of two trajectories 1 and 2 leading to the same scattering angle $\theta = \Theta(\lambda_1) = \Theta(\lambda_2)$ as illustrated in fig. 5.1.2 . The corresponding semiclassical cross-section

$$\frac{d\sigma_{sc}}{d\Omega} = \frac{d\sigma_{c\ell}^{(1)}}{d\Omega} + \frac{d\sigma_{c\ell}^{(2)}}{d\Omega} + 2\sqrt{\frac{d\sigma_{c\ell}^{(1)}}{d\Omega}\frac{d\sigma_{c\ell}^{(2)}}{d\Omega}} \; \cos(\beta_1 - \beta_2) \tag{5.1.25}$$

oscillates between $\left[(d\sigma_{c\ell}^{(1)}/d\Omega)^{1/2} + (d\sigma_{c\ell}^{(2)}/d\Omega)^{1/2}\right]^2$ and $\left[(d\sigma_{c\ell}^{(1)}/d\Omega)^{1/2} - (d\sigma_{c\ell}^{(2)}/d\Omega)^{1/2}\right]^2$ with period $\Delta\theta = 2\pi/|\lambda_1 \pm \lambda_2|$ (cf. eqs. (5.1.22) and (5.1.16)). Averaging over these oscillations gives the classical cross-section. The semiclassical approximation to this interference phenomenon is valid if the important regions around λ_1 and λ_2 with widths $\Delta\lambda_1$ and $\Delta\lambda_2$ do not overlap and if $\lambda_1, \lambda_2, \Delta\lambda_1, \Delta\lambda_2 \gg 1$. These conditions are violated for rainbow and glory scattering where two stationary points come close together.

5.1.2   Rainbow, glory and spiral scattering

The classical rainbow, glory and spiral scattering have been considered in section 2.3 . Because of interference effects we expect the semiclassical cross-sections to differ considerably from the corresponding classical expressions.

(i)   Rainbow scattering: According to our discussion above we have simple inter-ference between two trajectories as long as the scattering angle is far enough from the rainbow angle. If the scattering angle approaches the rainbow angle, the important regions around λ_1 and λ_2 begin to overlap. Therefore, the two contributions cannot be treated separately any more. In addition, $d^2\phi(\lambda)/d\lambda^2 \to 0$ and hence, third-order terms have to be included. Apart from this modification the rainbow scattering amplitude is calculated in the same way as in subsection 5.1.1. The analysis of the corresponding problem in optics was carried out by Airy. The cross-section is ex-

pressible in terms of Airy's integral, unless other deflection angles contribute
also:

$$\left(\frac{d\sigma}{d\Omega}\right)^{(r)}_{sc} = \frac{\lambda_r}{k^2 \sin\theta} \, 2\pi \, \frac{Ai^2(x)}{|\tilde{q}|^{2/3}} \qquad (5.1.26)$$

where $\tilde{q} = q/k^2$, with q of eq. (2.3.1). The function $Ai(x)$ is the Airy integral
defined by

$$Ai(x) \equiv \frac{1}{2\pi} \int_{-\infty}^{+\infty} \exp\left(ixu + \frac{1}{3}iu^3\right) du \qquad (5.1.27)$$

with $x = \tilde{q}^{-1/3}(\theta_r - \theta)$. The Airy integral is illustrated in fig. 5.1.3 . On the dark
side of the rainbow angle the intensity falls off more rapidly than exponentially.
On the bright side the intensity oscillates and corresponds under low resolution to
the classical prediction. The oscillation comes from the interference between the
two branches of the classical deflection function. The main shortcoming of the Airy
approximation is that the expressions (5.1.26) and (5.1.25) join onto each other
only if the deflection function is symmetric around λ_r and, therefore, is accurately

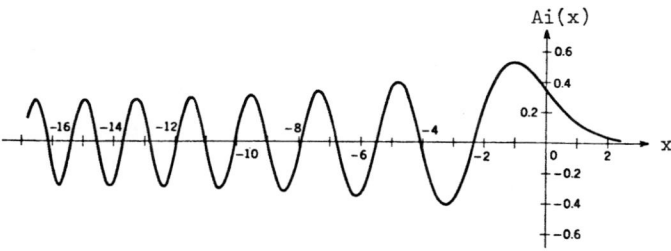

Fig. 5.1.3 The Airy integral Ai(x). (From [NE 66]).

given by the expansion (2.3.1). This is avoided by the uniform approximation [BE66,72].
Fig. 5.1.4 illustrates the modification of the classical rainbow scattering due to
quantal effects obtained in the uniform approximation [SI 73] .

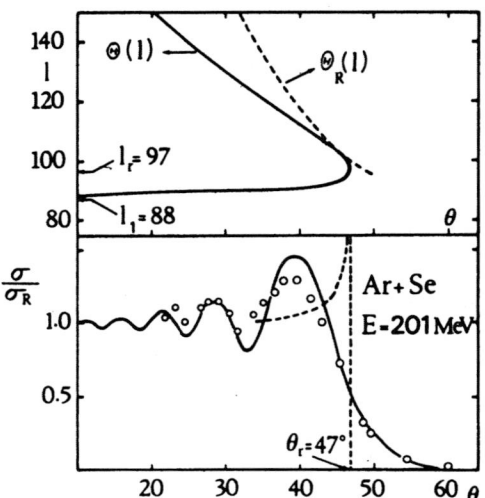

Fig. 5.1.4   Rainbow scattering for Ar+Se in the classical (---)
and the semiclassical approximation (—). $\Theta(\ell)$ is a
parametrized deflection function (parameters are
Θ_r, ℓ_r and ℓ_1). The Coulomb deflection function $\Theta_R(\ell)$
is shown for comparison. (From [SI 73]).

(ii)   Glory scattering: For the forward glory scattering treated in subsection 2.3.2, the stationary points at $+\Theta$ and $-\Theta$ come close together. Since $\Theta \approx 0$, we have to use the asymptotic expression (5.1.18) for $P_\ell(\cos \Theta)$. With $\tilde{\zeta} = \zeta/k$ (ζ is defined in eq. (2.3.3) and $\tilde{\zeta} \lambda_{glory} \gg 1$ we find in analogy to eqs. (5.1.19) to (5.1.23)

$$\left(\frac{d\sigma}{d\Omega}\right)_{sc}^{glory} = \frac{\lambda_{glory}^2}{k^2 |\tilde{\zeta}|} \ 2\pi \ J_o^2 \left(\lambda_{glory} \sin \Theta \right)$$
(5.1.28)

for the semiclassical glory cross-section. Oscillations can be ascribed to the Legendre polynomial $P_{\ell_{glory}}$ (cos Θ)   which corresponds to the glory λ value.

(iii)   Spiral scattering (orbiting) occurs in the classical scattering theory (cf. subsection 2.3.3) when the effective potential for the radial motion exhibits, for some angular momentum ℓ_{spiral} ,   a maximum which is equal to the available energy. Near this potential barrier $k \approx 0$ and hence, according to the conditions (5.1.9) the semiclassical approximation is violated. In most problems involving orbiting, it is necessary to take the barrier penetrability and resonance effects from virtual bound states into account.

5.1.3   Application and generalization of the semiclassical approximation

The semiclassical approximation described in this section, is expected to be applicable if the potential does not change appreciably within a wave length. This condition is very well satisfied in atomic and molecular collisions, but in heavy-ion collisions this is true only for large energies and/or very heavy nuclei. In the following we discuss two applications of the semiclassical theory: (i)   The ion-atom scattering $K^+(0.172eV)$ + Ar and (ii) the nucleus-nucleus scattering ^{16}O (60 MeV) + ^{58}Ni.

(i) For the scattering of an ion by an atom the interaction potential is well
approximated by a Lennard-Jones potential which is the sum of a long-ranged attractive part $-a/r^6$ and a repulsive core b/r^{12} . Generally speaking, many
ℓ-values contribute in the ion-atom scattering. But still, at some angles quantal
effects persist. However, since many partial waves contribute it is difficult to observe such quantum effects consisting in rapid oscillations of the cross-section as
function of energy and angle. In addition it is difficult to separate experimentally
the elastic from inelastic scattering. A survey of the scattering phenomena is given
in fig. 5.1.5 in an energy vs. angular momentum plot. The diagram illustrates the

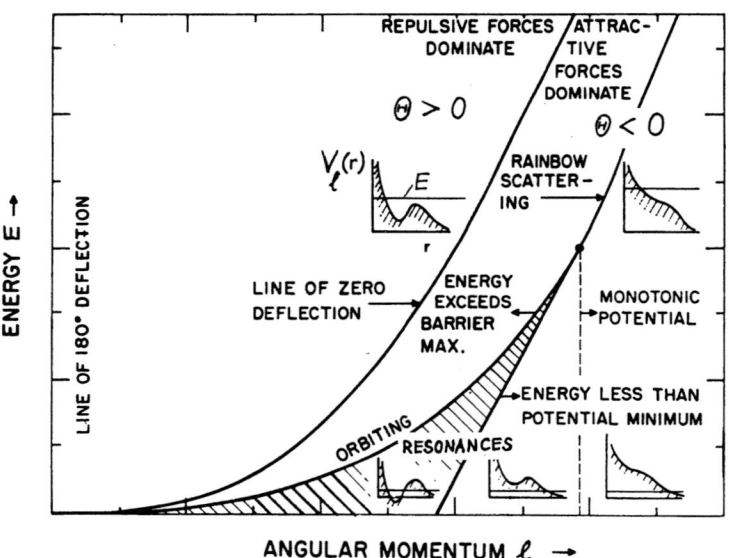

Fig. 5.1.5 Topology of ion-atom scattering in the energy vs.
angular momentum plane. (From [FO 59]).
The figure also applies to HI scattering.

places in (E,ℓ) space where rainbow, spiral and resonance scattering occur. Consider
for example a fixed energy (horizontal line in fig. 5.1.5) and let the angular
momentum (impact parameter) vary from infinity to zero. At first the deflection is
small and negative (net attraction) and becomes increasingly more negative because
of increasing attraction. The deflection angle then becomes either singular (orbiting)
or reaches a finite minimum (rainbow scattering). For decreasing ℓ the short-range
repulsion becomes important. The deflection angle increases and passes through zero
which corresponds to equal net attraction and repulsion along the trajectory. Finally,
$\Theta(\ell) \to \pi$ for $\ell \to 0$. Two typical deflection functions are shown in fig. 5.1.6.

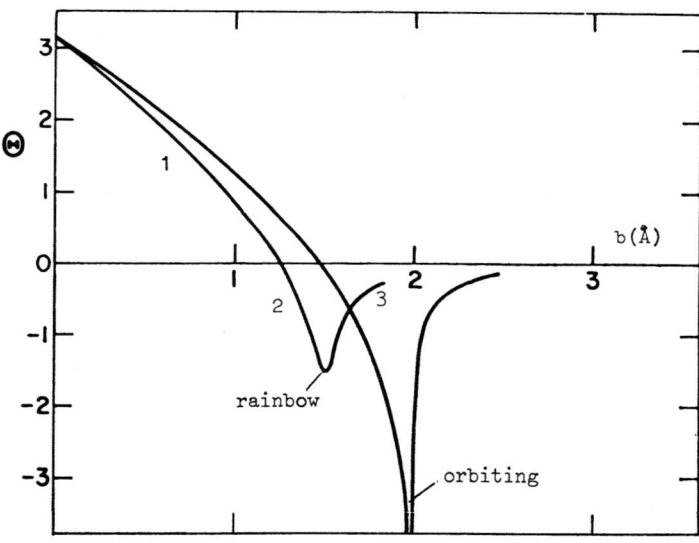

Fig. 5.1.6 Typical deflection functions for ion-atom scattering.
(From $\left[\text{FO } 59\right]$).

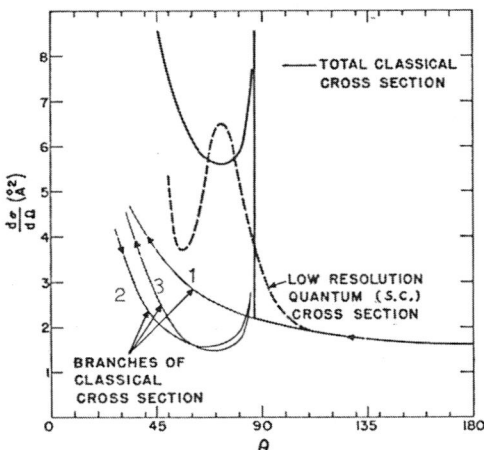

Fig. 5.1.7 Classical and averaged semiclassical cross-section
for K$^+$ on Ar. (From [FO 59]).

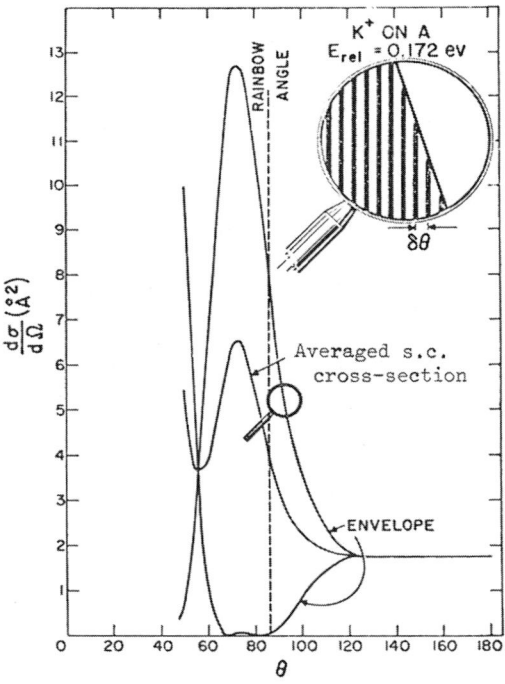

Fig. 5.1.8 Semiclassical cross-section for K$^+$ on Ar.
(From [FO 59]).

In fig. 5.1.7 the classical cross-sections for the scattering of K^+ ions by Ar atoms at $E_{c.m.}$ = 0.172 eV is shown. The corresponding deflection function is that of fig. 5.1.6 with a rainbow minimum. Between 0 and Θ_r three branches labelled 1, 2 and 3 in fig. 5.1.6 contribute to the classical cross-section. From Θ_r to π only the branch 1 contributes. The semiclassical cross-section, illustrated in fig. 5.1.8 shows strong oscillations due to the interference between the contributions from different branches. In the rainbow region the distance between neighbouring maxima or minima is 1.5° .

(ii) We return to the scattering of ^{16}O + ^{58}Ni at E_{lab} = 60 MeV which has been discussed in the framework of the classical approximation at the end of section 2.3 . In fig. 5.1.9 the semiclassical cross-section is compared with the classical and the quantum-mechanical cross-sections using potential II of [BR 72a] . The semiclassical cross-section (curve 1) results from the interference of different (up to three) branches of the deflection function and agrees remarkably well with the quantum-mechanical cross-section. A comparison with eq. (2.4.14) shows, indeed, that

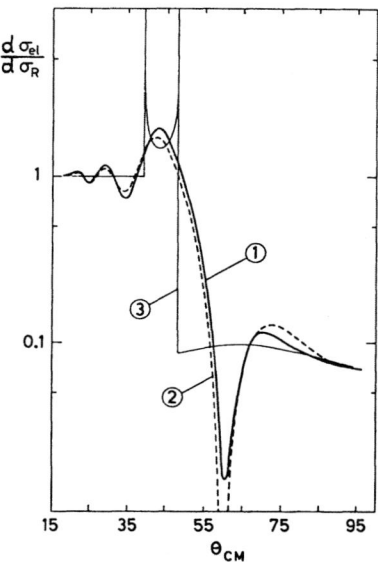

Fig. 5.1.9 Semiclassical (curve 1), quantum-mechanical (curve 2) and classical (curve 3) cross-section for $^{16}O(E_{lab}$ = 60 MeV) $+^{58}Ni$ for potential II (V_o = 7.0 MeV, a = 0.6 fm, r_o = 1.51 fm). (From [MA 73]).

$|\text{grad } \chi|^2 \approx 0.1 \ll 1$ and we expect the semiclassical approximation to be rather well justified.

Absorption may be taken into account by calculating the imaginary part of the phase shifts, eq. (5.1.13), in first order of the imaginary part $W(r)$ of the potential (weak absorption limit). The result is an attenuation factor in the semi-classical scattering amplitude (5.1.21), cf. [BR 74]

$$f_{sc}^{j}(\theta) = \left[f_{sc}^{j}(\theta) \right]_{no\,absorption} exp \left\{ \frac{2m}{\hbar^2} \int_{a_j}^{\infty} dr \, \frac{W(r)}{k_j(r)} \right\} . \qquad (5.1.29)$$

We consider the deflection function II of fig. 2.3.2 . For $b < 5$ fm the trajectories leading to larger angles correspond to smaller impact parameters and hence, smaller turning distances a_j . Therefore, we expect for $\theta_{cm} > 50^{\circ}$ the absorption to increase with increasing scattering angle. Calculating the cross-section with absorption according to eq. (5.1.29) for the example shown in fig. 5.1.9 one finds [BR 74] for $\theta_{cm} \leq 60^{\circ}$ a curve very similar to the two curves 1 and 2 of the figure. For $\theta_{cm} > 60^{\circ}$ it continues to decrease monotonically in sharp contrast to the steep rise of curves 1 and 2 .

The procedure (5.1.29) of including absorption excludes in principle the refractive and diffractive effects originating from the imaginary part of the inter-action. As discussed by several authors, e.g. [HA 74a, AU 75], diffraction is ex-pected to be important in all HI reactions. In order to treat also the effects from the complex interaction we have to generalize the simple semiclassical approximation by including the complex solutions (complex trajectories) of the classical equation of motion [KN74,75, MA75, KO 75]. The applicability of the semiclassical theory is thus extended to $|\text{grad } \chi| \approx 1$. Rainbow scattering and diffraction (Fresnel and Fraunhofer) are seen to emerge from this approach. Although these methods may become powerful tools in the theory of HI reactions, we do not treat them here. We discuss the diffraction phenomena in the sharp cut-off diffraction model (section 5.3) and the delicate mixing of refraction and diffraction in the smooth cut-off model (section 5.4). Before coming to these points and in order to become familiar with the general behaviour of the phase shifts in HI scattering, we treat in the follow-ing section 5.2 the optical model.

5.2   Optical model for elastic scattering

According to section 4.1, eqs. (4.1.8) and (4.1.9), the optical model potential is defined in such a way as to produce the energy averaged S or T matrix. But this definition determines the optical model potential not uniquely for the following reasons: The elastic channel is uniquely defined only in the asymptotic region. We can add any exponentially decreasing function of the relative coordinate. This might be physically reasonable, if account should be taken of nuclear polarization. Hence, $V_{opt}(r)$ depends on the choice of the channel wave-function. It is rather obscure how to define the elastic channel wave-function for two heavy ions. More formally the phase shifts do not define $V_{opt}(r)$ uniquely. It is known [NE 66, ch. 20] that for ℓ fixed, $\phi_\ell(E)$ (if known for all E) determines uniquely a local (spherically symmetric) potential or, for E fixed, all $\phi_\ell(E)$ determine uniquely a local potential if that potential is known not to have any bound states. However, we do not have any reason to assume that $V_{opt}(r)$ should be local. Hence, the optical potential is ambiguous. The ambiguity in the definition of the optical potential manifests itself in well-known difficulties: For weak absorption, a discrete family of local potentials exists which essentially differ from each other by the number of modes of the corresponding wave functions in the internal region. For strong absorption, the family of equivalent potentials even becomes continuous, because it is only the surface properties of $V_{opt}(r)$ that matter. This indicates that optical-model calculations are largely redundant, especially for heavy ions.

5.2.1   Parameters of the optical-model potential and redundancies

The optical-model potential is usually parametrized in terms of the sum of a real and an imaginary Woods-Saxon potential,

$$V_{opt}(r) = - \frac{V_o}{1 + exp \frac{r - R_r}{a_r}} - i \frac{W_o}{1 + exp \frac{r - R_i}{a_i}} \quad . \tag{5.2.1}$$

The volume absorption in eq. (5.2.1) is sometimes replaced or modified by a surface absorption. Reasonable optical-model potentials have also been obtained from the folding procedure [DO 75] , cf. section 2.2 . Let us consider HI scattering which is characterized by strong absorption. One may ask the question: What is essential

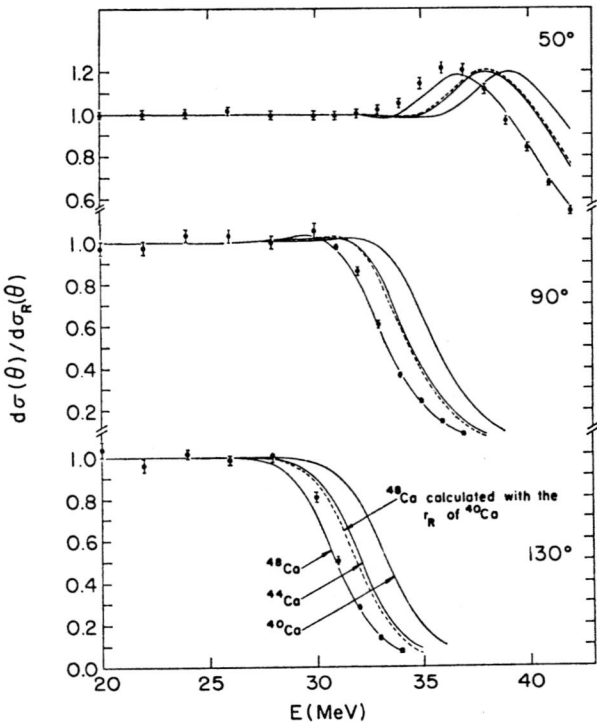

Fig. 5.2.1 Measurements of $d\sigma/d\sigma_{Ruth}$ for $^{16}O + ^{48}Ca$ and the optical model fits to these data as well as for $^{16}O + ^{40}Ca$ and $^{16}O + ^{44}Ca$. In order to indicate the sensitivity to variations in the optical-potential parameters the $^{16}O + ^{48}Ca$ scattering has also been calculated with the parameters for $^{16}O + ^{40}Ca$. (From [BE 71]) .

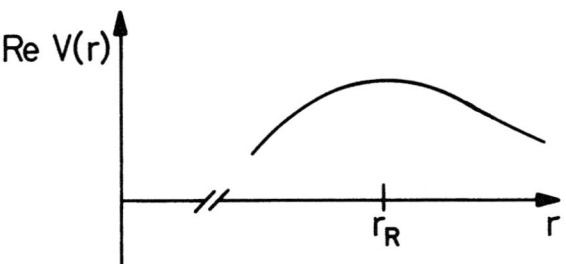

Fig. 5.2.2 The dependence of the real part of $V(r) = V_{Clb}(r) + V_{opt}(r)$ around the Coulomb barrier $(r = r_R)$.

in the optical model or more precisely, what are the essential parameters of $V_{opt}(r)$? This problem has been investigated by Bertin et al. [BE 71] and Eisen and Vager [EI 72] . Bertin et al. measure $d\sigma/d\sigma_{Ruth}$ for fixed scattering angle as function of the c.m. energy E . The result is shown in fig. 5.2.1 for the scattering $^{16}O + {^{48}}Ca$. In the optical-model fits Bertin et al. assume $a_r = a_i = \bar{a}$ and $R_r = R_i = \bar{R}$ so that they have four fit parameters. The fits turn out to be insensitive to W_o and, therefore, $W_o = -10$ MeV is assumed. Furthermore, the fits turn out to be sensitive only to two combinations of the remaining parameters $\bar{R}$, $\bar{a}$ and V_o, namely, the position r_R ("Rutherford radius") of the maximum of $ReV(r)$ with $V(r) = V_{Clb}(r) + V_{opt}(r)$ and the value $ReV(r_R)$ of the potential at this point (see fig. 5.2.2). This can be explained as follows. We notice that we are still way outside of the nuclear potential, so that $exp[(r_R - \bar{R})/\bar{a}] \approx 30$. The Rutherford radius r_R is considerably larger than the radius $\bar{R}$ of the interaction potential (5.2.1). Expanding $V(r)$ around $r = r_R$ we find in second order

$$Re\, V(r) \approx Re\, V(r_R) + \frac{1}{2} \frac{Z_1 Z_2 e^2}{r_R^3} \left[2 - \left(1 - \frac{r_R\, Re\, V(r_R)}{Z_1 Z_2 e^2}\right)^{-1} \right] (r - r_R)^2 \qquad (5.2.2)$$

which is completely determined by $ReV(r_R)$ and r_R .

Using a rather different approach, Eisen and Vager [EI 72] lend support to the result of Bertin et al. that only two parameters are essential in the optical-model potential. They use the incoming wave boundary condition (IWB) model of Rawitscher [RA 66] to simulate the strong absorption in the internal region. This model differs from the usual optical model in setting a boundary condition on the partial waves at a distance $d = r_R - 3\bar{a}$ which is smaller than the Rutherford radius. At this boundary, one imposes the condition that the partial waves have the form of incoming waves. This incoming wave boundary condition has been originally introduced by Blatt and Weisskopf [BL 63, p.351] . The IWB implies the absence of reflected waves from the nuclear interior due to strong absorption. It turns out that the results are not very sensitive to the value of d if d is chosen within the range between $r_R - 4\bar{a}$ and $r_R - 2\bar{a}$. Outside the boundary (i.e. for $r > d$) only a real optical potential is used which is approximated by $-\tilde{V}_o exp(-r/\bar{a})$. Taking

the Coulomb interaction to be that of two point charges and introducing the Rutherford radius r_R as defined above, one finds that $\widetilde{V}_o$ can be expressed in terms of
r_R and $\bar{a}$,

$$\widetilde{V}_o = \frac{Z_1 Z_2 e^2}{r_R^2} \, \bar{a} \, exp \, (r_R / \bar{a})$$
(5.2.3)

Again only two parameters, the Rutherford radius r_R and the diffuseness $\bar{a}$, are
needed for describing the scattering. The resulting fit parameters r_R = 9.31 fm,
$\bar{a}$ = 0.49 fm for 42 MeV ^{16}O on ^{48}Ca turn out to be close to the optical model parameters $(r_R$ = 9.25 fm, $\bar{a}$ = 0.52 fm). Fig. 5.2.3 shows a comparison of the reduced
scattering matrix for the IWB and the optical model. Unfortunately, this comparison
is _not_ made for the _same_ potential parameters (as it should), but rather for the
best-fit parameters just given, which differ in the two cases. To our knowledge this
is the only available study of the significant parameters in the optical model. Analytical studies of this problem do not seem to exist.

5.2.2   The elastic scattering of ^{16}O + ^{16}O

The elastic scattering of ^{16}O on ^{16}O has been the subject of extensive theoretical and experimental studies. What do we hope to understand from an analysis of
this reaction ?   In the regions where E fixed, $\theta > \theta_{gr}$ (cf. fig. 3.4.2) or θ
fixed, $E > E_{gr}$ (cf. fig. 3.4.3) we expect the strongest dependence on the details
of the heavy-ion interaction, and structure effects may come into play. It turns out
that in these domains, the scattering of ^{16}O on ^{16}O reveals the possible existence
of "orbiting" resonances. The relation between such resonances and a Regge pole
model is discussed in section 5.5 . In the following discussion we closely follow
the paper by Gobbi et al. [GO 73] . The authors study the scattering of identical
particles, because the number of ℓ-waves is reduced.

The measured excitation functions at various angles are shown in fig. 5.2.4
together with various optical-model fits in the energy range 10 MeV < E_{CM} < 40 MeV.
Typically, these excitation functions show a gross structure (typical width 2 MeV)
of rather equidistant maxima and minima. The intermediate structure (typical width

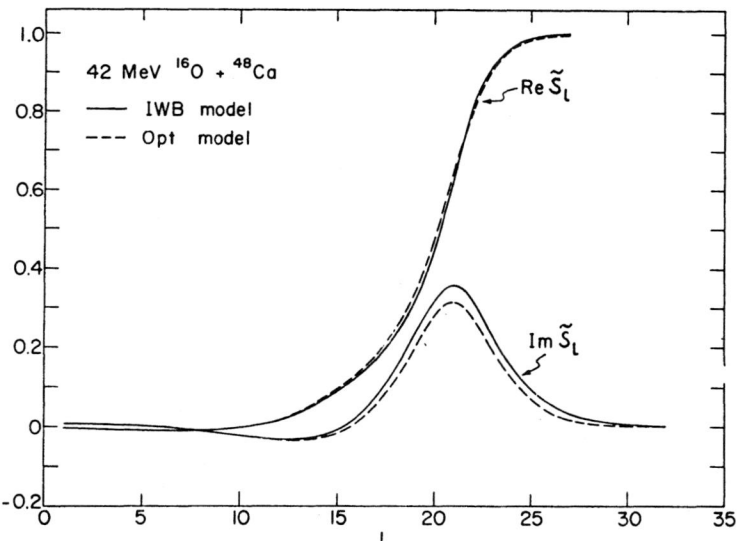

Fig. 5.2.3 Reduced scattering matrix $\tilde{S}_\ell = exp\,(2i\delta_\ell)$, cf. eqs.(5.3.25)
and (5.3.26) ,   extracted in the IWB and the optical
model for $^{16}O(42$ MeV$) + {}^{48}Ca$. These coefficients are
calculated from the best-fit parameters. (From [EI 72]) .

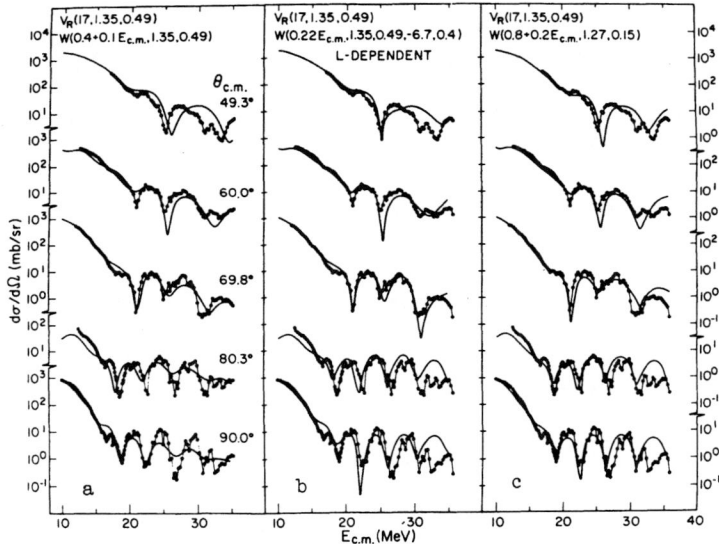

Fig. 5.2.4 Excitation functions for $^{16}O + {}^{16}O$ and different
optical-model fits. (From [GO 73]).

200 keV) is ascribed to compound nucleus effects. Good fits to the measured excitation functions (without the 200 keV fluctuations) have been obtained from an optical-model potential with V_o = 17 MeV, R_r = 6.8 fm, a_r = 0.49 fm, W_o = 0.8 + 0.2 E, R_i = 6.4 fm, a_i = o.15 fm (see fig.2.5.4c). This potential is characterized by a small value of V_o (shallow potential), by a strong increase of absorption with increasing energy, by a small radius but rather sharp surface for the absorptive part which is probably characteristic for closed shell nuclei. The fit which is obtained with this potential is reasonable for all excitation functions except for the last peak at θ = 90° .

Fig. 5.2.5 is an optical-model representation of the $^{16}O + ^{16}O$ scattering. At the extreme left the real potentials for the different partial waves (including V_{Clb} and centrifugal barrier) are shown. Next are shown the real part of the nuclear phase shifts[†] δ_ℓ and the reflection coefficients $\eta_\ell \equiv |exp(2i\delta_\ell)|$, the calculated 90° differential excitation function, and finally an energy-angle map of the calculated elastic scattering. We point out the main properties of this figure:

(i)   At fixed energy the real parts of the nuclear phase shifts δ_ℓ are negative ("reflection") for small ℓ-values and become positive for large ℓ-values.

(ii)   At any given energy there are only 2 or 3 angular momenta ℓ with η_ℓ significantly different from 1 or 0 . Moreover, so long as $n_\ell \approx 1$, the real part of the phase shift is very nearly equal to the Coulomb phase shift. This means that the entire structure at backward angles $\theta > \theta_{gr}$ is due to a few angular momenta. For example, $n_{18} \approx 1$ for E $\leq$ 25 MeV, = 0.5 at E = 27 MeV (dip) and $\approx$ 0 at 29 MeV. At certain energies the angular distribution at backward angles turns out to be characterized by a simple $|P_\ell(\cos\theta)|^2$. In particular, the maxima in the 90° excitation function at 21, 25, 29, ... MeV correspond to ℓ = 14, 16, 18, ...

To obtain a more complete understanding of the behaviour of individual partial waves, we consider the Argand diagrams for ℓ = 16, 18, and 20 as shown in fig. 5.2.6 . In diagrams (a), (b) and (c) the function $exp(2i\delta_\ell)$ is displayed, whereas in diagrams (d), (e) and (f) the partial amplitude $(2\eta)^{-1}(2\ell+1) exp[2i(\sigma_\ell - \sigma_o)][1 - exp(2i\delta_\ell)]$ is plotted. At the energy where a

---

[†] For a precise definition, see page 128 .

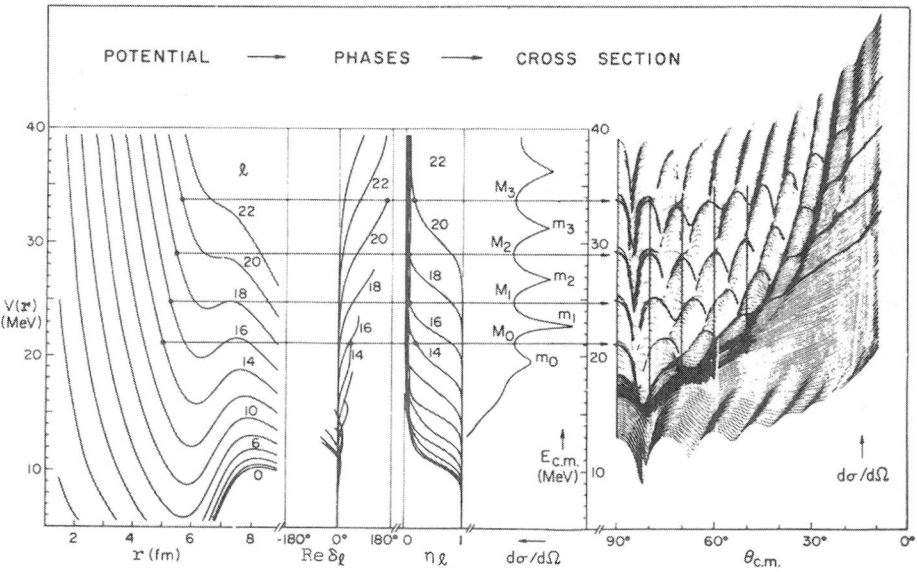

Fig. 5.2.5 Optical-model representation of the $^{16}O + ^{16}O$
elastic scattering. (From [GO 73]).

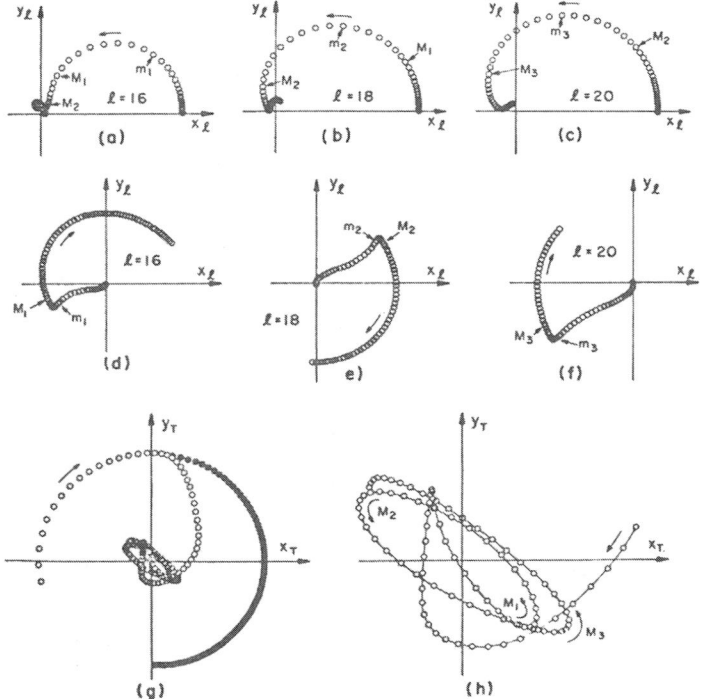

Fig. 5.2.6 Argand diagrams for the elastic $^{16}O + ^{16}O$
scattering. (From [GO 73]).

given partial wave begins to contribute significantly to the cross-section a fast
rise in the partial amplitude and almost no change in phase is observed. With
increasing energy a region occurs wherein the partial amplitude is stationary because
the Coulomb and nuclear phase shifts move in opposite directions. To illustrate the
origin of maxima and minima we consider the partial wave $\ell = 18$, i.e. fig. 5.2.6e.
For $E < 25$ MeV this partial wave contributes rather little to the total scattering amplitude. As it increases, it interferes destructively with the background
giving the minimum m_2 at 27 MeV. Then it remains virtually constant while the background changes. This gives the maximum M_2 at 29 MeV. The last two diagrams (g)
and (h) in fig. 5.2.6 display the total scattering amplitude at $\theta = 90°$. At low
energies the Coulomb term dominates. However, at higher energies the interference of
different partial waves reduces the cross-section. The origin of the oscillations in
the 90° differential cross-section are particularly evident in (h) which is an enlargement of the central region of (g).

For a better understanding of the mechanism generating the shape displayed
by the Argand diagrams (a), (b) and (c) of fig. 5.2.6, the magnitude $W = cW_o$ of
the imaginary potential has been systematically decreased in order to magnify the
shape effects caused by the real potential. The results obtained for $\ell = 18$ are
given in fig. 5.2.7 . With decreasing absorption the curves tend to the outer
circle which indicates a resonance. It is instructive to discuss in this connection
the form of the real potential and the behaviour of the radial wave-function for
this partial wave. Fig. 5.2.8 displays the energy variation of the radial wave-
function for $\ell = 18$. The increase of the radial wave-function just above the
barrier indicates the resonance phenomenon which is displayed in fig. 5.2.7 .

We have presented here some details from the study of Gobbi et al. [GO 73]
concerning the optical-model analysis of the excitation function and the angular
distributions in the elastic scattering of ^{16}O on ^{16}O . It was possible to arrive
at reasonable optical-model parameters and to understand the gross structure observed in the excitation function. This gross structure is attributed to the
existence of orbiting resonances, i.e. resonances due to two or three neighbouring

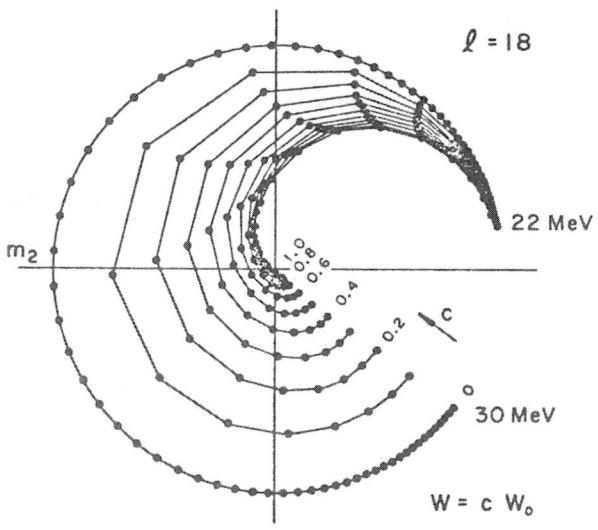

Fig. 5.2.7 Argand diagram displaying the energy dependence of
the scattering function for ℓ = 18 for variable
strength c of the imaginary potential. (From [GO 73]).

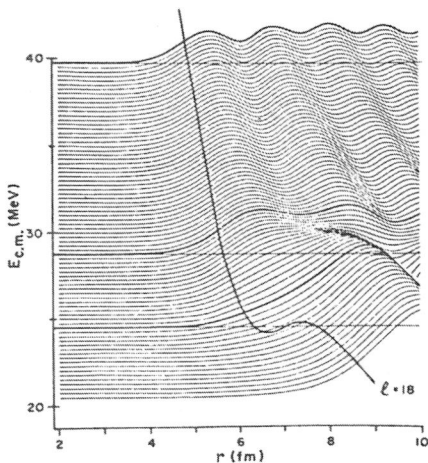

Fig. 5.2.8 Radial wave-function (ℓ = 18). The dark solid line
represents the real potential for ℓ = 18. (From [GO 73]).

partial waves which are strongly influenced by the onset of the nuclear attraction
at the surface. A beautiful description of such surface waves can be given within
the Regge-pole approximation which is discussed below in section 5.5 .

It might be expected from fig. 5.2.5 that the structures in the excitation
function are damped out for energies > 40 MeV, where the potential for the relevant
ℓ-waves has no minimum any more. This is not true both experimentally and theoretical-
ly [BR 73] . In fact, it has been shown by Halbert et al. [HA 74b] that the oscillations are present experimentally at least up to 90 MeV. Although the combined
potentials no longer have a minumum, they still seem to affect the phase shifts
in a manner similar to the one discussed in connection with figs. 5.2.5 and 5.2.6 .
The optical-model potential of Gobbi et al. [GO 73] which we have discussed here,
gives maxima and minima at approximately correct positions, but the magnitude at the
peaks is about 40 times too large. The origin of this discrepancy is expected to be
due to an energy dependence of the optical potential which is stronger than that
assumed by Gobbi et al. .

5.3   Sharp cut-off diffraction model

As we have seen in sections 5.1 and 5.2 , the elastic scattering of heavy
ions is characterized by the strong Coulomb repulsion and by strong absorption once
the two fragments start to overlap. In addition, the nuclear attraction may produce
specific effects in the cross-section, described as rainbow scattering (cf.section
5.1) or potential resonances (cf. section 5.2) . In this section we neglect the
nuclear attraction and discuss the diffraction pattern which is produced by strong
Coulomb repulsion and strong absorption. We do this by using not the full quantum-
mechanical description (this would involve a numerical solution of the Schrödinger
equation of a strongly absorptive optical-model potential), but rather the analogy
to optical diffraction theory to obtain the solutions in analytical form. This
enables us to display explicitly the typical features (Fraunhofer and Fresnel
diffraction) of the diffraction process.

We make the following assumptions

(i)   The absorption is very strong, so that the real part of the interaction potential can be neglected. This is a good approximation for very heavy ions, but
becomes worse for light projectiles such as p,d and α-particles. (cf. fig. 2.4.1).

(ii)   The absorbing potential has a sharp surface, i.e. we neglect the diffuseness.

(iii) The energy is large or, more precisely, kR >> 1 or ℓ_{gr} >> 1 .

It can be shown that under these assumptions, solutions of the Schrödinger equation
behave in a way known from optics and can be described in terms of (Fresnel)
diffraction patterns. The assumption (ii) allows us to obtain the diffraction pattern
directly from the optical analogy. In a more adequate treatment (cf. subsection 5.3.5),
this assumption is replaced by a sharp cut-off in the transmission coefficients T_ℓ
leading essentially to the same diffraction pattern. In describing these diffraction
phenomena in HI scattering we follow closely the paper of Frahn [FR 72] .

5.3.1   Qualitative considerations

As shown below, the solution of the Schrödinger equation reduces to a
diffraction problem in the limit

$$\ell_{gr} = ka_{sc} \gg 1 \tag{5.3.1}$$

where a_{sc} characterizes the linear dimensions of the scatterer. In this limit, the
problem becomes completely analogous to the diffraction of light. From optics we know
two types of diffraction patterns, characterized by the quantity $ka_{sc}\left(\frac{a_{sc}}{d}\right)$ with
d = min (d_1, d_2) where d_1 and d_2 are the distances of the scatterer from the
source and the point of observation, respectively. We distinguish the Fraunhofer
diffraction for which

$$ka_{sc}\left(\frac{a_{sc}}{d}\right) \ll 1 \tag{5.3.2}$$

and the Fresnel diffraction for which

$$ka_{sc}\left(\frac{a_{sc}}{d}\right) \gtrsim 1 . \tag{5.3.3}$$

Thus, Fraunhofer diffraction occurs when both the source point and the observation point are effectively at infinity, and Fresnel diffraction when either the source or the observation point or both are at a finite distance from the object. In optics, one usually has Fresnel diffraction, because the point of observation is at a finite distance behind the scatterer. We shall see that in HI scattering we are concerned with the converse situation where the (virtual) source is close to the scatterer. Both situations, observation close to object and source close to object, give for the same geometrical parameters the same diffraction pattern, because of the reciprocity theorem by Helmholtz [BO 65] .

Diffraction by a half-plane is the simplest of the very few diffraction problems that can be solved exactly in closed form. A convex aperture of arbitrary shape can be approximated by a family of half-planes tangential to the rim, if the smallest radius of curvature along the rim satisfies the inequalities (5.3.1) and (5.3.3). This makes it possible to study the diffraction pattern in the vicinity of the shadow cone. For convex bodies (instead of the two-dimensional objects considered so far) the diffraction pattern is completely determined by the shadow line which separates the illuminated part from the dark part of the body's surface (fig. 5.3.1). Therefore, if the conditions above are satisfied for the shadow line, all convex bodies having the same shadow line produce identical diffraction patterns and can, therefore, be replaced by an opaque disc D bounded by this shadow line.

5.3.2   Review of Fresnel diffraction

We study the diffraction by a half-plane [SO 59, § 38] . We consider an opaque screen which is infinitely extended and infinitely thin (idealization), and a linear polarized incident plane electromagnetic wave with electrical field vector $\vec{E}$ parallel to the edge of the screen, see fig. 5.3.2 . The plane wave is incident under the angle α . We are interested in the transition between the shadow region III and the lit region II, separated by the thick line in fig. 5.3.2. Introducing the plane polar coordinates r, φ , z and writing

$$\vec{E} = \vec{e}_z \, u \, (r, \varphi) \qquad (5.3.4)$$

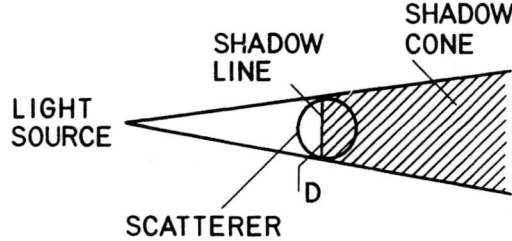

Fig. 5.3.1 Diffraction by a convex body and equivalent disc D

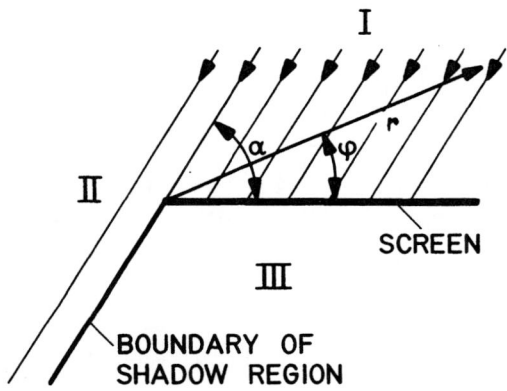

Fig. 5.3.2 Diffraction by a half-plane. (From [SO 59]).

we reduce the problem to finding the scalar function u as function of r, φ .
Since $\vec{E}_{tang}$ = 0 on the screen we have

$$u (r, \varphi = 0) = u (r, \varphi = 2\pi) = 0 \tag{5.3.5}$$

and since $\vec{E}$ is a solution of Maxwell's equation, $u(r, \varphi)$ has to be the solution
of the wave equation

$$\Delta u + k^2 u = 0 \tag{5.3.6}$$

outside the screen. In addition to the boundary condition (5.3.5), we have the"out-
going-wave" condition

$$u (r, \varphi) \overset{r \to \infty}{\sim} u_o (r, \varphi) + v (r, \varphi) \tag{5.3.7}$$

with $\quad u_o (r, \varphi) = \begin{cases} A \exp \left[-ikr \cos (\varphi - \alpha) \right] & \text{in } \mathrm{II} \\ 0 & \text{in } \mathrm{III} \end{cases} \tag{5.3.8}$

and $v \sim$ outgoing cylindrical waves only. To make the solution unique one also needs
Meixner's edge condition $\lim_{r \to 0} (\vec{r} \cdot \text{grad } u) = 0$.

The approximation (5.3.5) is good for sufficiently small wave lengths. The
nature of the material and the way in which the electromagnetic wave partly penetrates
into the screen are irrelevant. These phenomena only influence the vicinity of the
screen, which is explicitly excluded by (5.3.1) in the diffraction theory. The
solutions obtained here are also relevant quantum-mechanically: Eq. (5.3.6) corresponds to the time-independent Schrödinger equation of a free particle. Therefore,
in the diffraction limit the problem of the scattering of a particle by an opaque
half-plane is just the same as for light.

The result derived by Sommerfeld [SO 59] with the help of Green's function
and somewhat tricky contour integrals, leads to the well-known Fresnel formulas:
The total intensity $|u|^2$ behind the screen is given by

$$\left| \frac{u}{A} \right|^2 = \frac{1}{2} \left\{ \left[\frac{1}{2} + C(x) \right]^2 + \left[\frac{1}{2} + S(x) \right]^2 \right\} \tag{5.3.9}$$

where C(x) and S(x) are the real and imaginary parts of the Fresnel integral

$$F(x) = \int_0^x d\tau \exp(i\pi\tau^2/2) = C(x) + i S(x) \tag{5.3.10}$$

and $x = 2\sqrt{kr/\pi} \sin\frac{\tilde{\delta}}{2}$ with $\tilde{\delta} = \pi - (\varphi - \alpha)$ which is positive in the lit region
and negative in the shadow region. Right at the boundary (x=0) we have $|u/A|^2 = 1/4$.
The absolute value of the amplitude is shown in fig. 5.3.3 . The value $|u/A| = 1/2$
at the classical shadow edge $\tilde{\delta} = 0$ has been qualitatively explained by Blair [BL 54] :
According to Huygens' principle there is half of the amplitude of the outgoing waves
missing for $\tilde{\delta} = 0$.

5.3.3   Coulomb scattering and Fresnel diffraction

We shall establish here a simple physical relation between scattering in a
Coulomb field and Fresnel diffraction. Remember (cf. section 2.1) the elastic scattering of a projectile by the Coulomb field of a spherical target nucleus. The Coulomb

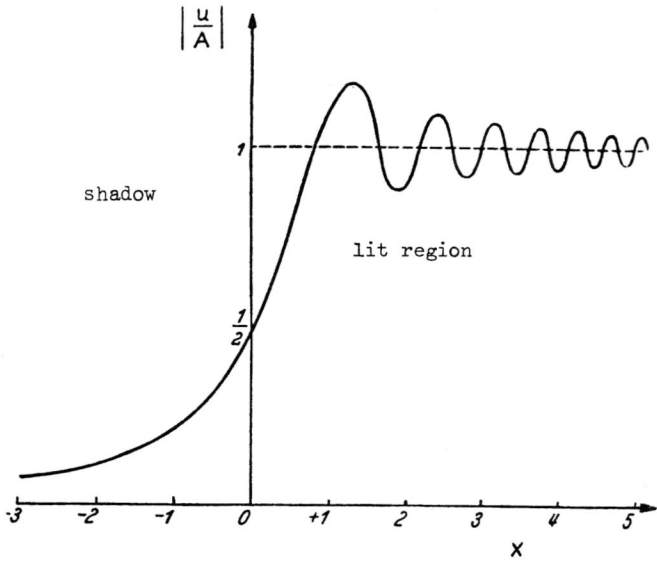

Fig. 5.3.3 Diffraction amplitude for Fresnel diffraction by
a half-plane. (From [SO 59]   ).

field distorts the impinging plane wave in the way indicated in fig. 5.3.4 . This Coulomb distortion acts as if a source for particles without Coulomb interaction were located at a finite distance d in front of the scatterer. The distance d depends on the scattering angle Θ . From simple geometrical considerations we find that

$$d = b / \sin \theta \overset{(2.1.7)}{=\!=} \frac{\alpha_c}{4E} \frac{1}{\sin^2 \frac{\theta}{2}} \qquad\qquad (5.3.11)$$

for pure Coulomb scattering. If the nucleus is strongly absorbing (sharp surface), the scattering situation corresponds to Fresnel diffraction by a black sphere with a point source located at the distance d corresponding to the grazing angle Θ_{gr} . In order that one gets a clear diffraction pattern corresponding to a point source, $\Delta d / d_{gr} \ll 1$ has to be fulfilled, where Δd is given by the change of d over the characteristic angle $\Delta\Theta$ of the interference pattern. From eq.(5.3.9) we find $\Delta\theta = \sqrt{2\pi \sin \theta_{gr} / l_{gr}}$. With $\Delta d / d_{gr} = ctg \left(\frac{\theta_{gr}}{2}\right)\Delta\theta$ from eq. (5.3.11) the condition $\Delta d / d_{gr} \ll 1$ becomes

$$l_{gr} \gg \frac{1}{\pi} ctg \frac{\theta_{gr}}{2} \cos^2 \frac{\theta_{gr}}{2} \qquad\qquad (5.3.12)$$

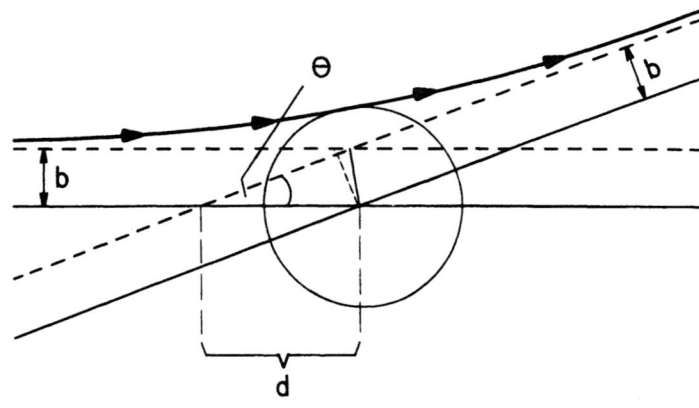

Fig. 5.3.4 The diverging lense action of the Coulomb field
and the virtual source points.

which is usually fulfilled for diffraction scattering, condition (5.3.1), if θ_{gr} is not too small.

From the geometry of Coulomb scattering (fig.5.3.4), it is seen that the radius a_{sc} of the equivalent disc (or that of the complementary circular aperture) is the impact parameter b_{gr} of the grazing trajectory, not (as one might expect) the interaction radius R . The diffraction condition (5.3.1) now reads

$$\ell_{gr} = \frac{1}{\hbar} P_\infty b_{gr} \overset{(2.4.17)}{=} \eta \, ctg \, \frac{\theta_{gr}}{2} \gg 1 \tag{5.3.13}$$

and according to eqs. (5.3.2) and (5.3.3), we can distinguish the two diffraction types:

(i)   Fresnel scattering for

$$k a_{sc} \left(\frac{a_{sc}}{d} \right) \xrightarrow{(5.3.11)} \ell_{gr} \frac{b_{gr}}{b_{gr}/\sin\theta_{gr}} = \ell_{gr} \sin\theta_{gr} \gtrsim 1 \; ; \tag{5.3.14a}$$

(ii)   Fraunhofer scattering for

$$\ell_{gr} \sin\theta_{gr} \ll 1 \; . \tag{5.3.14b}$$

If $\sin\theta_{gr} \ll 1$ we have $d \gg b_{gr}$ ("long distance condition", cf. fig. 5.3.4) and $\theta_{gr} \ll 1$ ($\pi - \theta_{gr} \ll 1$ is excluded because of condition (5.3.13)). In this case the distinction between Fresnel and Fraunhofer scattering is completely determined by the value of the Sommerfeld parameter

$$\eta \approx \frac{1}{2} \ell_{gr} \theta_{gr} \; . \tag{5.3.15}$$

Ideal Fraunhofer and Fresnel diffraction patterns occur in the following limits:

a) Fraunhofer limit: $\ell_{gr} \to \infty$, $\eta/\ell_{gr} \to 0$ so that $b_{gr} \to R$ and $\theta_{gr} \to 0$. This limit corresponds to high energy ($E_{CM} \to \infty$) where the Coulomb interaction is negligible. The differential cross-section takes the familiar form

$$\frac{d\sigma}{d\Omega} = R^2 \left[\mathcal{J}_1(kR\theta) / \theta \right]^2 \tag{5.3.16}$$

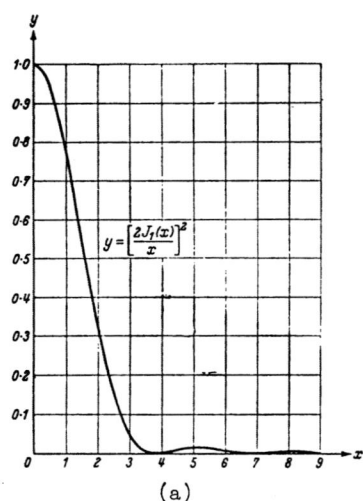

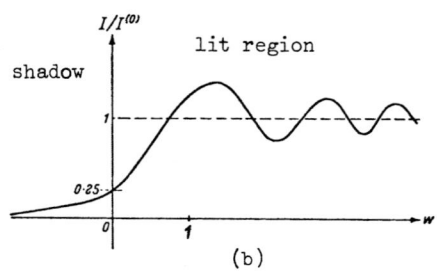

(b)

Reproduced with permission from M.Born and E.Wolf
Principles of Optics (Pergamon, Oxford, 1965)

(a)

Fig. 5.3.5   Ideal Fraunhofer (a) and Fresnel (b) diffraction pattern.
(From [BO 65]).

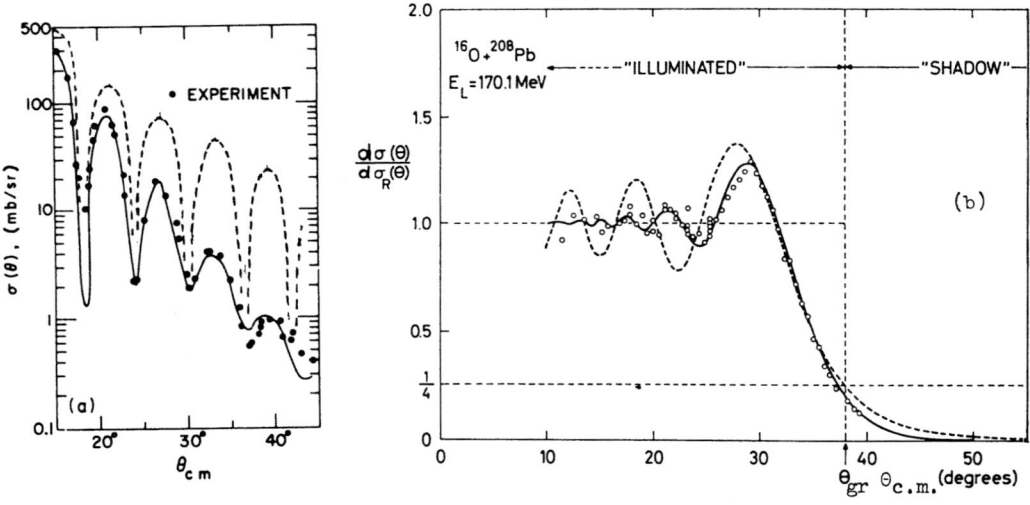

Fig. 5.3.6   Example of (a) Fraunhofer diffraction pattern for
$^{16}O + ^{12}C$ at E_{CM} = 72 MeV [HI 64] and (b) Fresnel
diffraction pattern for $^{16}O + ^{208}Pb$ at E_{CM} = 158 MeV
[FR 72] . The broken curves are calculated for $^{12}C + ^{16}O$
from eq. (5.3.16) with R = 6.45 fm and for $^{16}O + ^{208}Pb$
from eq. (5.3.18) with η = 31.2 and ℓ_{gr} = 90.4 . The
solid curves contain modifications due to the finite
surface thickness and the real part of the phase shifts.
(From [SI 74] and [FR 72]) .

for the scattering by a perfectly absorbing sharp-edged sphere. The angular distribution is shown in fig. 5.3.5a . This pattern is affected by the nuclear potential
in the surface region: The diffuseness of the absorptive potential suppresses the
diffraction maxima at larger angles. The real part fills in the minima. An example
is shown in fig. 5.3.6a .

b) Fresnel limit: $\ell_{gr} \to \infty$, $\eta/\ell_{gr} = tg(\theta_{gr}/2) \to$ const. Thus the Fresnel limit
corresponds to high energies $(\ell_{gr} \to \infty)$ and strong Coulomb interaction $(\eta \to \infty)$
at a fixed ratio of the center-of-mass energy to the Coulomb barrier, because of
relation (3.3.2). The diffraction pattern for the ideal Fresnel limit is described
by eqs. (5.3.9) and (5.3.10) with the substitutions

$$kr \to kd_{gr} = \ell_{gr} / \sin\theta_{gr}$$
$$\tilde{\delta} \to \theta_{gr} - \theta \tag{5.3.17}$$

which give

$$\frac{d\sigma}{d\sigma_{Ruth}} = \left|\frac{u}{A}\right|^2 = \frac{1}{2}\left\{\left[\frac{1}{2} + C(x)\right]^2 + \left[\frac{1}{2} + S(x)\right]^2\right\},$$
$$x = 2\sqrt{\frac{\ell_{gr}}{\pi \sin\theta_{gr}}} \sin\frac{\theta_{gr} - \theta}{2}. \tag{5.3.18}$$

The diffraction pattern is shown in fig. 5.3.5b where we use x instead of

$$\theta \approx \theta_{gr} - x\sqrt{\frac{\pi \sin\theta_{gr}}{\ell_{gr}}}. \tag{5.3.19}$$

For $\theta < \theta_{gr}$ the cross section shows the Fresnel oscillations. This is illustrated
in fig. 5.3.6b . Notice that the observed grazing angle is usually different from
the expression (3.3.3) used throughout this section. Due to the attraction by the
nuclear interaction, the grazing trajectory is bent forward, so that the experimentally determined grazing angle is somewhat smaller (cf. fig.3.4.2)

Significant deviations from this pattern due to nuclear structure effects in
HI scattering are expected at backward angles where diffraction gives a small cross-
section. The actual damping of the Fresnel oscillations in the forward direction
(cf. fig.5.3.6b) is due to the diffuseness of the absorbing potential and maybe also
due to the real part of the potential and to the "extension" of the virtual source

as discussed after eq. (5.3.11).

5.3.4 Frahn's diffraction diagram

We have seen that in the diffraction conditions (5.3.13) and (5.3.14) two parameters ℓ_{gr} and $p = \ell_{gr} \sin \Theta_{gr}$ are needed to specify a scattering situation with regard to its diffraction characteristics. This enables us to classify all scattering situations of charged particles above the Coulomb barrier by means of a two-dimensional diagram. It is convenient to choose as parameters the ratio

$$h = \frac{E_{CM}}{V_{CB}} = \frac{1}{2} \frac{kR}{2\eta} = \frac{1}{2} \left(1 + \frac{1}{\sin(\Theta_{gr}/2)} \right) \tag{5.3.20}$$

which is completely determined by the grazing angle, and the Sommerfeld parameter

$$\eta = \frac{Z_1 Z_2 e^2}{\hbar v} \quad . \tag{5.3.21}$$

The parameters ℓ_{gr} and p are related to the new parameters h and η by

$$\ell_{gr} = 2\eta h \sqrt{1 - \frac{1}{h}} \tag{5.3.22}$$

$$p = 2\eta \left(1 - \frac{1}{(2h-1)^2} \right) \tag{5.3.23}$$

The plot of η versus h gives the "diffraction diagram" shown in fig. 5.3.7 . The total area is divided somewhat arbitrarily into three main regions:

   (I) $\ell_{gr} \lesssim 10$, nondiffractive scattering ,

   (II) $\ell_{gr} \gtrsim 10$ and $p \lesssim 0.1$, Fraunhofer scattering ,

   (III) $\ell_{gr} \gtrsim 10$ and $p \gtrsim 1$, Fresnel scattering .

Because of the gradual changes in diffraction behaviour these lines should not be regarded as sharply defined boundaries. Moreover, they apply only to pure Coulomb scattering. The triangular areas A, B and C indicate the range of η and p which is obtained for the scattering of composite particles (deuterons to ^{238}U ions) by different target nuclei at $E_{Lab} = 10,20$ and 100 MeV/nucleon, respectively; the line labelled P corresponds to proton-nucleus scattering at 10 MeV. Therefore, in HI scattering, we expect to be mostly concerned with Fresnel scattering. In the

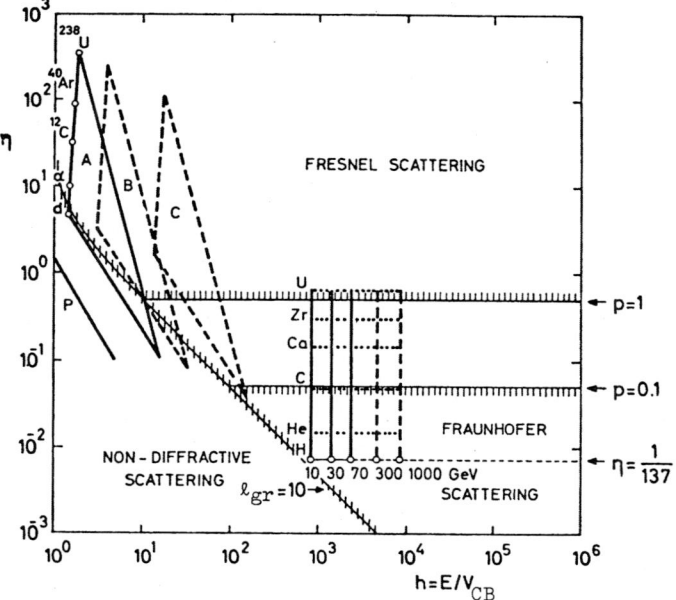

Fig. 5.3.7 Diffraction diagram for scattering of charged particles
above the Coulomb barrier. (From [FR 72]). Note that
in actual HI collisions the transition from Fresnel to
Fraunhofer diffraction is observed at larger p values
(around p = 10 , cf.fig.5.3.6a)

relativistic region v → c the Sommerfeld parameter attains the value

$$\eta \;\rightarrow\; \frac{Z_1 Z_2 e^2}{\hbar c} \;=\; \frac{Z_1 Z_2}{137} \;,\qquad\qquad (5.3.24)$$

i.e., η approaches a constant as function of h . The vertical lines represent the
scattering of protons (and other singly-charged hadrons) by protons and complex
nuclei at E_{Lab} between 10 and 1000 GeV. The dotted horizontal line at η = α =
1/137 is the trajectory for hadron-hadron scattering in the asymptotic energy
region.

It is possible that a given point in the diffraction diagram represents two
or more physically different scattering conditions. We call such scattering situations homomorphic. Since the diffraction pattern depends only on ℓ_{gr} and Θ_{gr} in
the way indicated in eq. (5.3.18), it should be the same if ℓ_{gr}, p or η,h are

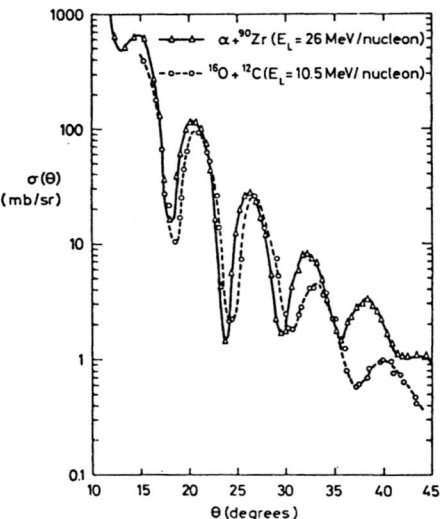

Fig. 5.3.8 Example of two nearly homomorphic diffraction patterns.
Solid curve: 104 MeV α on ^{90}Zr corresponding to
(η,h) = (2.46, 7.15). Dashed curve: 163 MeV ^{16}O on ^{12}C
corresponding to (η,h) = (2.33, 7.22) in the diffraction
diagram. (From [FR 72]).

the same. An example of a close, though not exact, homomorphism is shown in fig.5.3.8.

    Details of the diffraction diagram for the HI scattering are shown in
fig. 5.3.9   for E$_{Lab}$ = 10 MeV/nucleon: Curves are shown for constant values of the
diffraction parameters ℓ_{gr} and p . Solid lines represent the scattering of a given
projectile (identified by its symbols ^{A}X at the upper left) by a range of
target nuclei (labelled by their atomic number Z$_2$). For instance, the scattering
situations of fig. 5.3.6 are represented by the second and last point on the ^{16}O
curve.

    Let us finally discuss the connection between diffraction scattering and
classical scattering. The condition for the classical Coulomb scattering is
η^2 >> η^2$_{crit}$ (θ) with η$_{crit}$ (θ) defined by eq. (2.4.19). For θ ≈ θ$_{gr}$ =
const , i.e., h = const , we now consider the classical limit η → ∞ , i.e., ℓ_{gr}→ ∞
according to eq. (5.3.22). According to eq. (5.3.19) the diffraction pattern (see
fig. 5.3.10) becomes narrower and shrinks (but not uniformly) towards the classical
shadow boundary with increasing η ∝ ℓ_{gr} until it becomes indistinguishable from
the classical limit for η → ∞ .

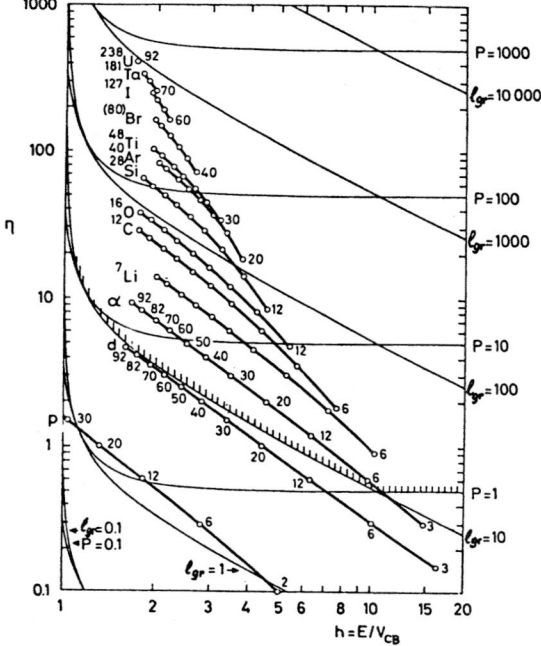

Fig. 5.3.9 Diffraction diagram for nucleus-nucleus scattering at E_{Lab} = 10 MeV/nucleon. (From [FR 72]).

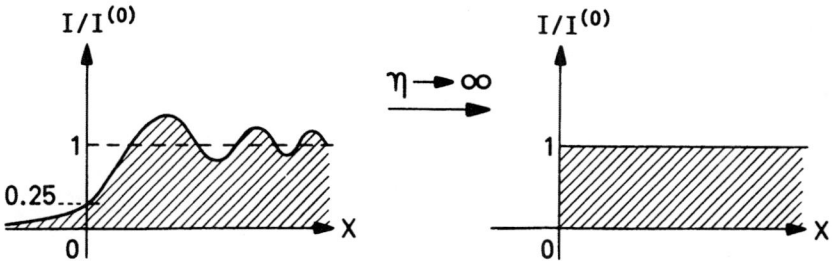

Fig. 5.3.10 Transition from Fresnel to classical scattering.

5.3.5   Frahn's derivation of Fresnel diffraction from the strong absorption model

We give this derivation here, because it casts additional light upon the physical assumptions used in a Fresnel description of èlastic HI scattering. The method of derivation is similar to the method used in the semiclassical approximation and in the smooth cut-off model.

Let us introduce in eq. (4.1.1) the difference $\delta_\ell = \phi_\ell - \sigma_\ell$ between the total phase shift ϕ_ℓ and the Coulomb phase shift σ_ℓ, hereafter referred to as nuclear phase shift. (Note that δ_ℓ differs from the phase shift derived from the nuclear potential alone!) Then the scattering amplitude takes the form $\left[\text{ME 61, eq. (XI.55)}\right]$

$$f(\theta) = f_{Clb}(\theta) + \frac{1}{2ik} \sum_{\ell=0}^{\infty} (2\ell+1)\, e^{2i\sigma_\ell} \left(e^{2i\delta_\ell} - 1 \right) P_\ell(\cos\theta) \qquad (5.3.25)$$

where $f_{Clb}(\theta)$ is the Coulomb scattering amplitùde. This equation is convenient for a Coulomb field plus a short-ranged potential because $\delta_\ell = 0$ for ℓ-values so high that the additional potential is not felt. Then, only $f_{Clb}(\theta)$ contains contributions from high partial waves.

Frahn introduceś four assumptions in eq. (5.3.25) in order to derive Fresnel diffraction from the strong absorption model.

(i)   In order to simulate the scattering by a strongly absorbing object he assumes that the reduced scattering matrix $\widetilde{S}_\ell$ has the form

$$\widetilde{S}_\ell \equiv \exp(2i\delta_\ell) = \begin{cases} 0 & \text{for } \ell \leq \ell_{gr} \\ 1 & \text{otherwise} \end{cases} \qquad (5.3.26)$$

which is known as the Blair model $\left[\text{BL 54}\right]$. Thus, the nuclear phase shift δ_ℓ is assumed to be zero for $\ell > \ell_{gr}$, The lower partial waves $\ell \leq \ell_{gr}$ are assumed to be completely absorbed. Note that the sum over ℓ in eq. (5.3.25) is now finite.

(ii)   The upper limit ℓ_{gr} is assumed to be large compared to 1. Since the main contribution to the sum comes from $\ell \lesssim \ell_{gr}$, we can use the asymptotic expression (5.1.17) for $P_\ell(\cos\theta)$ valid for large ℓ and $\sin\theta > 1/\ell$ and replace the sum over ℓ by an integral over $\lambda = \ell + \frac{1}{2}$.

(iii)   In order to get a simple analytic expression for σ_ℓ we assume $\eta \gg 1$ and use the corresponding asymptotic form. This assumption restricts the derivation to $p \gtrsim 10$ as given by eq. (5.3.23) and thus to the upper part of the diffraction diagram (fig. 3.5.7) excluding in particular the Fraunhofer scattering.

(iv)   The integral

$$f(\theta) - f_{C\ell b}(\theta) = \int d\lambda \mid f(\lambda) \mid e^{i\Phi(\lambda)}$$

(5.3.27)

which results from the assumptions (i) to (iii), is approximately evaluated by the
method of stationary phase, i.e., by expanding $\Phi(\lambda)$ around λ_0 (defined by
$(d\Phi/d\lambda)_{\lambda=\lambda_0} = 0$) up to second order. The result is

$$f(\theta) = f_{C\ell b}(\theta) + \frac{1}{2k} \sqrt{\frac{\eta}{\pi \sin^2\frac{\theta}{2}}} \exp\left(2i\sigma_0 - 2i\eta \ln\left[\sin\frac{\theta}{2}\right] + i\frac{\pi}{4}\right) \int_{\tau=-\infty}^{\tau=-x} d\lambda \exp\left(i\pi\tau^2/2\right)$$

(5.3.28)

where $\lambda_0 = \eta \, ctg \frac{\theta}{2}$, $\tau = \sqrt{\frac{2}{\pi\eta}} (\lambda - \lambda_0) \sin\frac{\theta}{2}$, $x = 2\sqrt{\frac{\Lambda}{\pi \sin\theta_{gr}}} \sin\frac{\theta_{gr}^c - \theta}{2}$
$\Lambda = \ell_{gr} + 1/2$ and θ_{gr}^c defined by $ctg\frac{\theta_{gr}^c}{2} = \Lambda/\eta$. With
$\left[\text{ME 61 , eq. (XI.33)}\right]$

$$f_{C\ell b}(\theta) = - \frac{\eta}{2k \sin^2\frac{\theta}{2}} \exp\left(2i\sigma_0 - 2i\eta \ln\left[\sin\frac{\theta}{2}\right]\right)$$

(5.3.29)

and $d\lambda = \sqrt{\frac{\pi\eta}{2}}\left(\sin\frac{\theta}{2}\right)^{-1} d\tau$ we get

$$f(\theta) = f_{C\ell b}(\theta)\left[1 + \frac{1}{\sqrt{2}} \exp\left(i\frac{\pi}{4}\right) \int_\infty^x d\tau \exp\left(-i\pi\tau^2/2\right)\right] .$$

(5.3.30)

Introducing the (real) Fresnel integrals $C(x)$ and $S(x)$ as in eq. (5.3.10) and
using $\int_\infty^0 d\tau \exp\left(-i\pi\tau^2/2\right) = -\frac{1-i}{2}$ we find for the scattering amplitude

$$f(\theta) = \frac{1}{2} f_{C\ell b}(\theta)\left\{1 + \left[C(x) + S(x)\right] + i\left[C(x) - S(x)\right]\right\}.$$

(5.3.31)

The cross section is given by

$$\frac{d\sigma(\theta)}{d\sigma_{Ruth}(\theta)} = \left|\frac{f(\theta)}{f_{C\ell b}(\theta)}\right|^2 = \frac{1}{2}\left\{\left[\frac{1}{2} + C(x)\right]^2 + \left[\frac{1}{2} + S(x)\right]^2\right\}$$

(5.3.32)

which is identical with the Fresnel formula given in eq. (5.3.9). In contrast to
the relation between the grazing angle and the grazing angular momentum given earlier,
see e.g. eq. (5.3.13) there occurs now $\Lambda = \ell_{gr} + \frac{1}{2}$ instead of ℓ_{gr} in this relation.

The replacement $\ell_{gr} \rightarrow \Lambda$ is a well-known modification arising in quantum mechanics. For $\Lambda \gg 1$ it is irrelevant.

The sharp cut-off limit discussed so far in this section seems to be rather realistic for HI reactions. For instance, reactions like ^{16}O + Ca and α+ Pb can be described by an optical potential with V = 100 MeV and W = 20 MeV at the center. For incident energies of about 25 MeV, one finds that the mean free path is ≈ 0.5 and 1 fm for s-wave oxygen and α-paricles, respectively [EI 72] . For such mean free paths which are small compared to nuclear dimensions one can use the sharp cut-off model as a first approximation. Corrections come from

(i)   the gradual onset of the absorption and the existence of a real potential
         and from

(ii)   nuclear structure effects, e.g. Regge poles and elastic transfer.

The latter features significantly affect the cross-section at large scattering angles. Since the diffraction pattern is practically zero at backward angles, such quantum effects are observable there. Some corrections to the sharp cut-off model (Blair's model) originating from (i) are described by the smooth cut-off model which is treated in the following section. Corrections due to Regge poles are discussed in section 5.5 .

5.3.6   Rainbow refraction versus Fresnel diffraction

The forward oscillations shown in figs. 5.1.4, 5.1.9 and 5.3.6b can be interpreted either as rainbow scattering or as Fresnel diffraction. From their generalized semiclassical approximation Knoll and Schaeffer [KN 75] conclude that these forward oscillations are essentially produced by the rainbow and, therefore, are due to refraction. This is concluded from the study of ^{16}O + ^{16}O scattering. In contradiction, Frahn [FR 75] concludes from the study of the smooth cut-off model (cf. the following section 5.4) that absorption and, therefore, diffraction is responsible for these oscillations. At present this question cannot be settled. It might be that both mechanisms are more or less important. According to the discussion of subsection 2.4.2 we expect a tendency towards a diffraction pattern for heavier colliding nuclei, because the (Coulomb) rainbow is becoming more and more masked by absorption. Some light is cast onto the interplay between rainbow refraction and Fresnel diffraction by the smooth cut-off model described below.

5.4   Smooth cut-off model

In view of its simplicity it is surprising that the sharp cut-off model is capable of describing essential features of HI scattering in terms of a single parameter $\Lambda = \ell_{gr} + \frac{1}{2}$. One expects that its shortcomings can easily be removed by modifying the idealized assumption (5.3.26). This has been done by several authors $\left[\text{GR 60, MC 60, EL. 61} \right]$ in the smooth cut-off model by introducing a gradual transition for the reflection coefficients $\eta_\ell = |S_\ell|$ from $\eta_\ell = 1$ for $\ell \gg \ell_{gr}$ to $\eta_\ell < 1$ for $\ell \approx 0$. In addition, the real part of the nuclear phase shifts δ_ℓ is taken to be different from zero. The qualitative behaviour of η_ℓ and Re δ_ℓ is illustrated in fig. 5.4.1 which is in agreement with the numerical values for $^{16}O + {}^{16}O$ scattering shown in fig. 5.2.5 . In analogy to the semiclassical deflection function (5.1.16) we define the quantal deflection function by

$$\Theta(\lambda) = 2 \frac{\partial}{\partial \lambda} \left[\sigma(\lambda) + \text{Re } \delta(\lambda) \right] = 2 \frac{\partial}{\partial \lambda} \text{Re } \phi(\lambda) \qquad (5.4.1)$$

where $\sigma(\lambda - \frac{1}{2}) \equiv \sigma_\ell$ and $\delta(\lambda - \frac{1}{2}) = \delta_\ell$ with $\lambda = \ell + \frac{1}{2}$ are the Coulomb and nuclear phase shift functions, respectively. Four qualitatively different cases can be distinguished (cf. fig. 5.4.1):

(i)   Weak absorption and weak nuclear attraction are characterized by $\delta(\lambda) \approx 0$, i.e. $\eta(\lambda) \approx 1$ (where η is the reflection coefficient) and Re $\delta(\lambda) \approx 0$, and by $\Theta(\lambda)$ close to the pure Coulomb deflection function.

(ii)   Weak absorption and strong nuclear attraction are characterized by $\eta(\lambda) \approx 1$ and Re $\delta(\lambda)$ significantly different from zero such that rainbow scattering (maxima or minima in $\Theta(\lambda)$) occur.

(iii)   Strong absorption and weak nuclear attraction are characterized by $\eta(\lambda) \approx 0$ for small λ (i.e. $\Delta/\Lambda \ll 1$, cf. fig. 5.2.5) and by Re $\delta(\lambda) \approx 0$ thus giving rise to a deflection function close to the Coulomb deflection function.

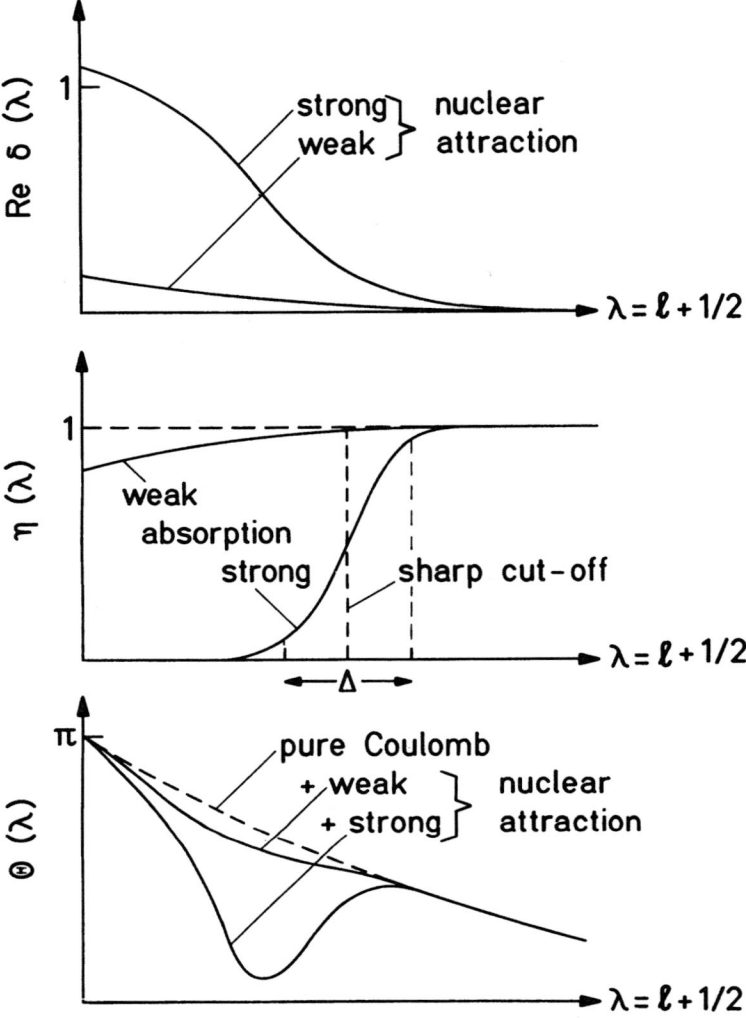

Fig. 5.4.1 Real part Re δ(λ) of the nuclear phase shift, reflection coef-
ficient η(λ) = |exp (2iδ(λ)) | and quantal deflection function
Θ(λ), cf. eq. (5.4.1), for strong and weak nuclear attraction
and absorption.

(iv)   Strong absorption and strong nuclear attraction are characterized by $\eta(\lambda) \approx 0$

for small λ (i.e. $\Delta/\Lambda << 1$, cf. fig. 5.2.5) and Re $\delta(\lambda)$ significantly dif-

ferent from zero such that rainbows arise in the quantal deflection function.

There are three kinds of parametrization possible: One can parametrize

either $\eta(\lambda) = |S(\lambda)|$ and Re $\phi(\lambda)$, or Re $S(\lambda)$ and ImS(λ), or the complex function

$S(\lambda)$. A convenient parametrization is that of McIntire et al. $\begin{bmatrix} MC\ 60 \end{bmatrix}$.

$$\eta(\lambda) = \frac{1}{1 + exp\ \dfrac{\Lambda - \lambda}{\Delta}}\quad ,$$

$$Re\,\phi(\lambda) = \sigma(\lambda) + \delta_o\left[1 - \eta(\lambda)\right] . \tag{5.4.2}$$

This parametrization is not consistent with general properties of the S-matrix. For

instance, the application of Regge pole theory to scattering by a complex optical-

model potential shows that, given $S(\lambda)$ as a function of λ, all poles of $S(\lambda)$

must lie in the upper half of the complex λ plane. This is not the case for eqs.

(5.4.2) and some other parametrizations used. It is not clear at present to what

extent this limits the applicability of these parametrizations in diffraction models.

Following Frahn $\begin{bmatrix} FR\ 73,\ 74 \end{bmatrix}$ we first give a general treatment of the smooth

cut-off model and discuss the results for the four cases (i) to (iv). The case of

strong absorption and weak nuclear attraction is discussed in some detail.

5.4.1   General treatment of the smooth cut-off model

The method for evaluating the scattering amplitude (4.1.1) in the smooth

cut-off model is a natural generalization of the treatment which leads to the semi-

classical approximation, cf. eqs. (5.1.19) to (5.1.22). With $S(\lambda) = \eta(\lambda)exp\left[2iRe\,\phi(\lambda)\right]$

we obtain instead of eqs. (5.1.19) and (5.1.20)

$$f(\theta) = f_+(\theta) - f_-(\theta) \tag{5.4.3}$$

$$f_\pm(\theta) = -\frac{1}{k\sqrt{2\pi\,sin\theta}} \int_{\frac{1}{2}}^{\infty} d\lambda\,\sqrt{\lambda}\;\eta(\lambda)\,exp\left[i\chi_\pm(\lambda)\right] \tag{5.4.4}$$

$$\chi_\pm(\lambda) = 2\,Re\,\phi(\lambda) \pm \lambda\theta \pm \frac{\pi}{4} \tag{5.4.5}$$

The main contributions to the integral in eq. (5.4.4) arise, as in eq. (5.1.19), from the vicinity of possible points of stationary phase of the integrand, and in addition from $\lambda \approx \Lambda$ (cf. eq. (5.4.2)) where $\eta(\lambda)$ varies rapidly.

In the case of weak absorption this variation is negligible and we are left with the semiclassical contribution from a stationary point λ_j with an additional attenuation factor $\eta(\lambda_j)$,

$$ f(\theta) \approx f_{sc}^{j}(\theta)\, \eta(\lambda_j) \tag{5.4.6} $$

which is of the same form as eq. (5.1.29). Thus, in the weak absorption limit, the smooth cut-off model reproduces qualitatively the results obtained in the semi-classical approximation. Note that a contribution from a stationary point $\left(\partial\, \mathcal{R}e\, \phi(\lambda)/\partial\lambda \right)_{\lambda=\lambda_s} = \theta/2$ in f_- corresponds to a classical trajectory leading to a positive deflection angle, whereas a contribution from a stationary point in f_+ $\left(\partial\, \mathcal{R}e\, \phi(\lambda)/\partial\lambda \right)_{\lambda=\lambda_s} = -\theta/2$ corresponds to a classical trajectory leading to a negative deflection angle.

For strong absorption the variation of $\eta(\lambda)$ around $\lambda = \Lambda$ becomes important. This contribution to the integrals (5.4.4) describes the diffraction due to absorption and does not have a semiclassical counterpart. Frahn [FR 73,74] has studied the strong absorption case within two limits. (1), in the limit of weak nuclear attraction which is treated in subsection 5.4.2 below. (2), the rainbow scattering in the sharp cut-off limit. Here he finds a Fraunhofer diffraction pattern superimposed upon a modified rainbow scattering.

5.4.2 Strong absorption model

We confine this discussion to the case of weak nuclear attraction. In this case (cf. fig. 5.4.1) there is only one point of stationary phase λ_s defined by $\Theta(\lambda_s) = \theta$. This contribution corresponds to a Coulomb trajectory which is slightly modified by the weak nuclear attraction. An additional diffractive contribution to both $f_+(\theta)$ and $f_-(\theta)$ arises from the vicinity of $\lambda = \Lambda$. The diffractive contribution to $f_-(\theta)$ is attributed to the diffraction from that side of the

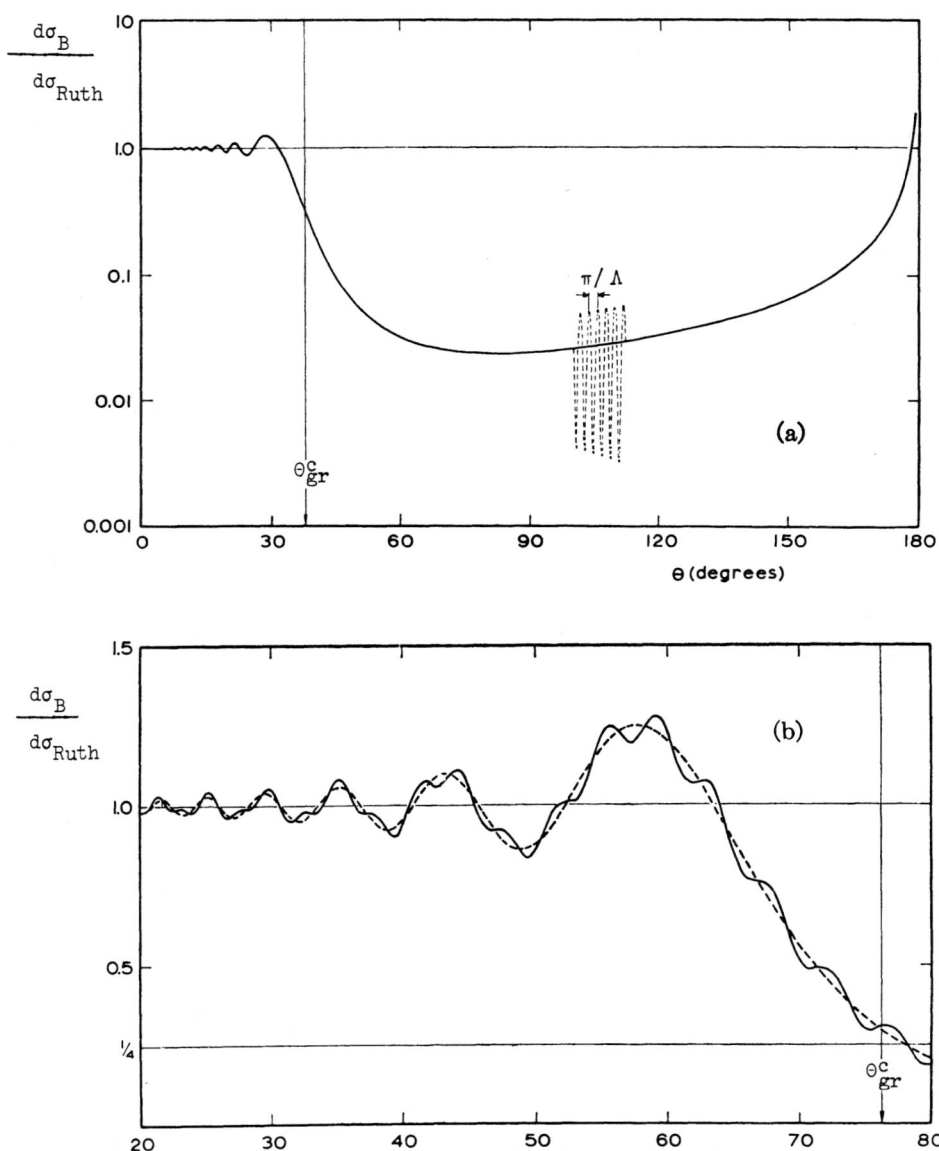

Fig. 5.4.2 (a) Blair model differential cross-section $d\sigma_B(\theta)/d\Omega$
divided by the Rutherford cross-section $d\sigma_{Ruth}(\theta)/d\Omega$ ($\eta = 31.07$,
$\ell_{gr} + 1/2 = 90$). The solid curve is obtained for $\theta > \theta^c_{gr}$
by averaging over the diffraction oscillations (indicated
by the dashed curve around $\theta = 100^o$), and for $\theta < \theta^c_{gr}$ by
averaging over the "small" oscillations. (b) $d\sigma_B(\theta)/d\sigma_{Ruth}(\theta)$
as function of θ for $\theta \leq \theta^c_{gr}$ ($\eta = 29.02$, $\ell_{gr} + 1/2 = 37$).
The dashed curve indicates the large oscillations. (From [FR 63]).

nucleus where the corresponding semiclassical trajectory passes. This roughly corresponds to scattering by a half plane, cf. fig. 5.3.2 . Therefore, $f_-(\theta)$ gives rise to Fresnel diffraction. The diffractive contribution to $f_+(\theta)$ is attributed to the diffraction from the other side of the nucleus. The interference of this contribution with $f_-(\theta)$ gives rise to a Fraunhofer diffraction pattern superimposed on the Fresnel pattern. Frahn [FR 73] has worked out general expressions for $f_\pm(\theta)$. We summarize the main results:

Fig. 5.4.2 shows the angular distribution for $\Delta = \text{Re}\delta(\lambda) = 0$ which defines the sharp cut-off model of section 5.3 . As compared to the ideal Fresnel diffraction pattern obtained in section 5.3 in the (classical) limit $\eta \gg 1$, we now get some qualitative changes. The point $d\sigma_B(\theta)/d\sigma_{Ruth}(\theta) = \frac{1}{4}$ is shifted to somewhat larger angles due to a correction of the order η^{-1} . For $\theta < \theta_{gr}^c$, there are in addition to the familiar "large" (Fresnel type) oscillations, "small" (Fraunhofer type) oscillations with period π/Λ . For $\theta > \theta_{gr}^c$, after the sharp drop-off, oscillations start where $d\sigma_B(\theta)/d\sigma_{Ruth}(\theta) \approx 1/\eta$, with amplitude rapidly increasing with angle and with a period π/Λ (asymptotically for $\theta - \theta_{gr}^c \gg 1$).

In the smooth cut-off model the diffraction pattern changes in the following way for $\text{Re}\delta(\lambda) = 0$:

(i)   The Fraunhofer type ("small") oscillations are strongly damped with increasing Δ . In the "dark" region for $\theta \gg \theta_c$ this damping depends for fixed Δ/Λ exponentially on the Sommerfeld parameter η if the parametrization (5.4.2) is used (Coulomb damping).

(ii)   The Fresnel amplitudes are much less damped, they are multiplied by the form factor $F(\Delta[\theta - \theta_{gr}^c])$ which is the Fourier transform of the absorptive shape function $D(\lambda) = d\eta(\lambda)/d\lambda$,

$$F(\Delta x) \equiv \int_{-\infty}^{+\infty} d\lambda \; D(\lambda) \; \exp\left[i \,(\lambda - \Lambda) x\right] \quad .$$

$$(5.4.11)$$

This function is $F(\Delta x) = \pi\Delta x \,(\sin h \,\pi x)^{-1}$ for the parametrization (5.4.2) which essentially decreases as $\exp(-\pi x)$ for increasing x .

These features are qualitatively unaffected by nonvanishing but small real nuclear phases.

5.5   Regge poles

We are interested in certain quantum-mechanical features which are not described by the diffraction models, or by the semiclassical approximation. Although these features are implicitly contained in some optical-model parametrization of the scattering process, it is useful to display them in a more explicit fashion. This helps in the interpretation not only of some elastic scattering data, but also of inelastic (transfer) reactions, and of the reaction mechanism itself. The mathematical tool introduced for this purpose is the Regge-pole description.

5.5.1   Definition and properties of Regge poles

We consider the radial equation for the partial wave $\varphi_\ell(k,r)$ with angular momentum ℓ for a real potential $V(r)$,

$$\left\{ - \frac{\hbar^2}{2m} \frac{d^2}{dr^2} + \frac{\hbar^2 \ell(\ell+1)}{2mr^2} + V(r) - E \right\} \varphi_\ell(k,r) = 0 \ , \tag{5.5.1}$$

and define solutions $\varphi_\ell(k,r)$ for complex values of ℓ by analytic continuation of eq. (5.5.1). This also defines a continuation of the phase shifts ϕ_ℓ. Putting $A_\ell = \sin\phi_\ell \, \exp(i\phi_\ell)$, we write the scattering amplitude $f(\theta)$ in the form

$$k f(\theta) = \sum_{\ell=0}^{\infty} (2\ell+1) A_\ell P_\ell (\cos\theta) \ . \tag{5.5.2}$$

Eq. (5.5.2) can be identically transformed into (Sommerfeld-Watson transformation)

$$k f(\theta) = \frac{1}{2\pi i} \oint d\ell \, (2\ell+1) \, P_\ell(-\cos\theta) \, A_\ell \, \frac{\pi}{\sin(\pi\ell)} \tag{5.5.3}$$

where the contour encloses the points $\ell = 0, +1, +2, +3, \ldots.$, but no other singularities of the analytic function $P_\ell A_\ell$. The residue of $\pi(\sin \pi\ell)^{-1}$ at the pole ℓ is $(-)^\ell$ which just cancels the same factor arising from $P_\ell(-\cos\theta) =$

$$= (-)^{\ell} P_{\ell} (\cos \theta).^{1)}$$

The contour of integration in eq. (5.5.3) is now deformed in such a way (see fig. 5.5.1) that it encloses the positive real ℓ-axis and the origin in a counterclockwise direction. For certain classes of potentials one can show [AL 65] that A_{ℓ} has only simple poles in the half plane $\text{Re}\,\ell > -1/2$, all of which lie above the real axis. Further deformation of the contour (dotted lines in fig. 5.5.1) leads to two types of contributions: The integral extended along the line $\ell = -\frac{1}{2} + i\alpha$ where $-\infty \leqslant \alpha \leqslant +\infty$, and a number of contributions arising from the poles $\ell = \alpha_n$ of A_{ℓ}. These are the Regge poles, defined as poles of the S-matrix obtained by analytic continuation in the angular momentum ℓ. We thus have

$$k f(\theta) = \frac{1}{2\pi i} \int\limits_{-1/2+i\infty}^{-1/2-i\infty} d\ell \, (2\ell+1) \, P_{\ell}(-\cos\theta) \, A_{\ell} \, \frac{\pi}{\sin(\pi\ell)}$$
$$+ \sum_n \beta_n \, P_{\alpha_n}(-\cos\theta) \, \frac{1}{\sin(\pi\alpha_n)} \qquad (5.5.4)$$

where β_n is $-(2\alpha_n+1) \cdot \pi$ times the residue of A_{ℓ} at $\ell = \alpha_n$. A formula like (5.5.4) is only useful if a few terms in the sum over n describe the behaviour of $kf(\theta)$ with good accuracy. One then has a Regge pole description of the scattering amplitude. For some potential models, this is the case at sufficiently high energy. (Hence the original introduction of the formula (5.5.3) by Sommerfeld in his studies of the scattering of electromagnetic waves at high energies, and the appeal of Regge poles to particle physicists).

---

1)

Note that the last equality holds only for integer values of ℓ. It is thus a matter of choice whether one uses $P_{\ell}(-\cos\theta)$ or $(-)^{\ell} P_{\ell}(\cos\theta)$ in eq. (5.5.3). The choice adopted here has the advantage that the representation of $f(\theta)$ obtained by subsequent manipulation holds for backward scattering angles, where quantal effects are strongest. A replacement of $P_{\ell}(-\cos\theta)$ by $(-)^{\ell} P_{\ell}(\cos\theta)$ would give a Regge pole description of forward scattering events. Because of the preponderance of diffraction scattering in this domain, the resulting formulas would not be very useful.

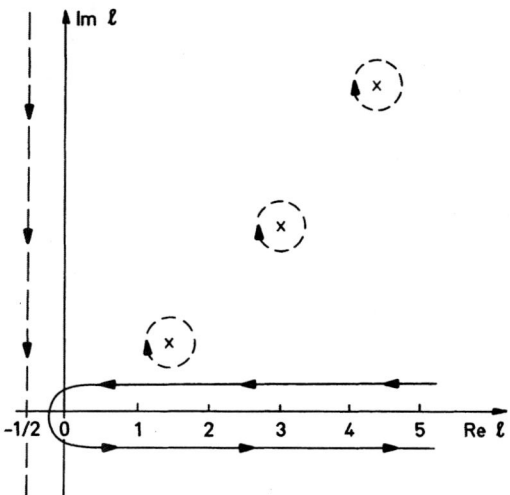

Fig. 5.5.1 Deformation of the contour of integration used to obtain
the Regge-pole representation of the scattering amplitude.
The crosses indicate Regge poles.

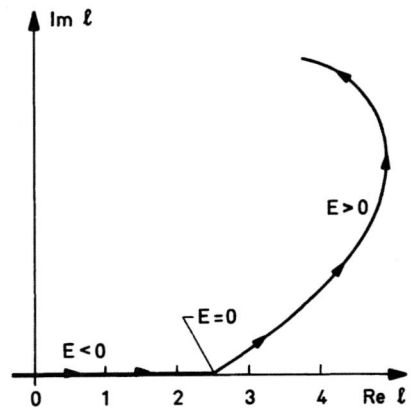

Fig. 5.5.2 A Regge trajectory for a Yukawa-type potential (schematic).

The location of a Regge pole depends upon energy, $\alpha_n = \alpha_n(E)$. The locus of $\alpha_n(E)$ in the complex ℓ-plane as E varies along the real E-axis is called a Regge trajectory. A typical trajectory is shown in fig. 5.5.2 . The arrow denotes the direction of increasing energy. The trajectory runs along the real axis so long as $E < 0$. It crosses integer values of ℓ, whenever E coincides with the energy of a bound state with orbital angular momentum ℓ. In this way various bound states of a potential are grouped into a "Regge family". This is the case, for instance, for the 1s, 1p, 1d, ... states which form the first Regge trajectory. The trajectory turns into the upper half of the complex plane at $E = 0$. For Re ℓ = integer, it describes a genuine resonance, if Im ℓ is not too large. To see this, let E_R be that value of E where Re ℓ is integer, and expand Re $\alpha_n(E)$ up to terms of first order in $E-E_R$. This gives, for a single Regge term in eq. (5.5.4)

$$-(-)^{Re\,\alpha_n(E_R)} \frac{\beta_n(E_R)\,P_{\alpha_n(E_R)}(-\cos\theta)}{\pi R\,(E-E_R+\frac{i}{2}\Gamma_R)} \quad \text{where} \quad R = Re\left\{ \frac{d}{dE}\alpha_n\Big|_{E_R} \right\} ,$$

$I = Im\,\alpha_n(E_R)$, and $\Gamma_R = 2I/R$. For trajectories of the type shown in fig. 5.5.2 we obviously have $R > 0$ and $I > 0$ so long as the pole is close to the real E-axis. Regge poles are obviously suited for the description of resonance phenomena where a sequence of partial waves, all with the same radial quantum number, i.e., a Regge trajectory, dominates the scattering. We have seen in subsection 5.2.2 that this is the case in the elastic scattering of ^{16}O by ^{16}O. The same idea has also been applied to α-scattering at backward angles.

We are thus led to the following picture $\left[CO\ 70,\ MC\ 71\right]$. The diffraction model (smooth cut-off model) gives a fair overall representation of the elastic scattering phase shifts, and its prediction may thus be identified with the integral and the far-lying pole contribution in eq. (5.5.4). The detailed behaviour of the phase shifts near the grazing angular momentum is, however, perhaps not well approximated. We attempt to improve the diffraction model by adding one or two Regge poles to the description. Similarly, one may proceed in modifying phase shifts obtained from optical-model potentials. The physical idea behind such a procedure is essentially

this. We found in section 5.2 on the optical model that the overall behaviour of the
phase shifts is quite well described by more or less all "reasonable" optical-model
potentials. Departures from this overall behaviour are due to specific ℓ-values
related to surface resonances. A given optical model may fail to give a Regge-type
behaviour, because it does not describe correctly details of the surface part of
the true potential. An improvement is possible by grafting on to this optical model
scattering amplitude one or two Regge poles. This procedure may be simpler than the
search for another optical-model potential.

To implement this picture, it is useful to realize which modifications in the
behaviour of Regge trajectories arise as one considers <u>complex</u> (rather than real)
potentials. For real potentials, unitarity implies that poles and zeros of $f(\theta)$
occur symmetrically about the real ℓ-axis, the poles occurring at $\operatorname{Im} \ell \geqslant 0$.
Roughly speaking, addition of an imaginary part to the potential means a broadening
of the resonances and a corresponding upward shift of the Regge trajectory (see
fig. 5.5.3). Because of these shifts the trajectory acquires a fairly large distance
from the real axis, and its influence upon $f(\theta)$ becomes weak. The dotted line in
fig. 5.5.3 gives the trajectory of the Regge zero. It also receives an upward shift

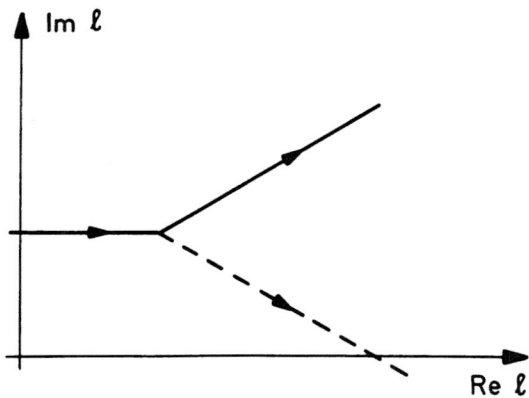

Fig. 5.5.3 Regge trajectories (full line) and Regge zeros (dotted
line) for a complex potential (schematic).

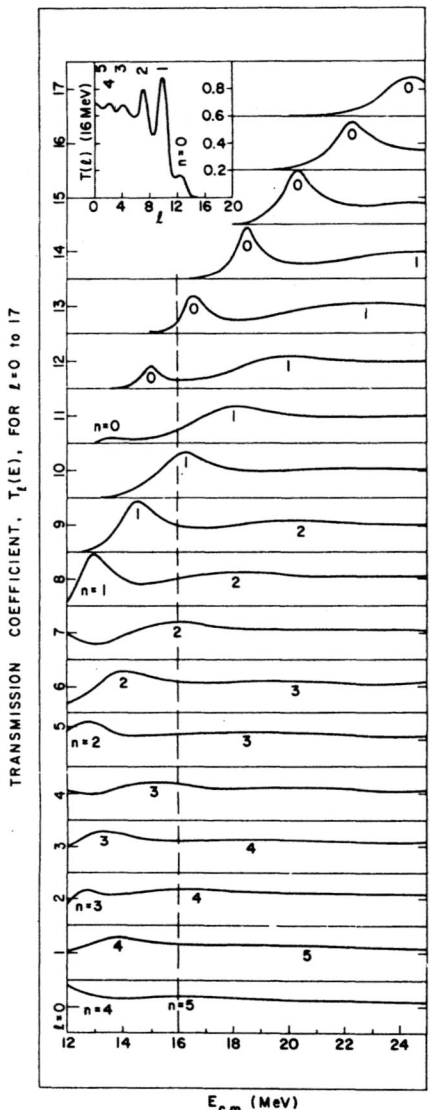

Fig. 5.5.4 Optical-model transmission coefficients T_ℓ as
functions of c.m. energy for angular momenta
$\ell = 0$ to $\ell = 17$. (From [MC 71]).

and, therefore, crosses the real ℓ-axis at some point. It appears that the main Regge behaviour of strongly absorbing potentials resides not in the pole trajectory, but in this crossover point of the zero trajectory.

This can be demonstrated by looking at the dependence on energy and angular momentum of the transmission coefficients $T_\ell = 1 - |S_\ell|^2$ calculated for the optical-model potential used by Maher et al. [MA 69] to fit their elastic $^{16}O + ^{16}O$ data. Fig. 5.5.4 taken from the work of McVoy [MC 71] shows this dependence. Note that maxima of T_ℓ correspond to minima of $|S_\ell|$, and that $T_\ell = 1$ at a zero of $|S_\ell|$. The figure shows maxima for T_ℓ labelled by their radial quantum numbers. These maxima move with increasing c.m. energy E_{CM} towards higher ℓ-values and finally disappear in the background. The $n = 0$ trajectory crosses the real ℓ-axis near $\ell = 15$, for instance. This is the last trajectory to move out of the potential well, it "leaves" the well at about $E_{CM} = 17$ MeV. Below the cross-over point, the resonance is mostly inelastic, its wave functions being strongly affected by absorption. Above this point, the resonance becomes more and more elastic: The increase in ℓ pushes the resonance towards the nuclear surface where absorption is reduced. Since the resonance energy increases more strongly than the angular momentum barrier, the elastic width increases. The total width increases monotonically with energy.

A more detailed investigation of Regge poles in optical-model potentials by Tamura and Wolter [TA 72] has shown that several differences exist between the trajectories shown in fig. 5.5.2 for real Yukawa-type potentials and those found for optical-model potentials of the Woods-Saxon type. (i) The latter are more or less straight lines and do not seem to show the backbending seen in fig. 5.5.2. (ii) Because of the behaviour of S_ℓ for large values of Im ℓ, the contour of the background integral cannot be chosen along a parallel to the imaginary axis. (iii) A single-pole (or single-zero) parametrization appears useful at sufficiently high values of E_{CM}, and applies only to the $n = 0$ trajectory, so that the other trajectories contribute to the smooth background. For $n > 0$, several trajectories are important in determining the behaviour of $f(\theta)$.

5.5.2 Diffraction model and Regge poles

McVoy [MC 71] suggested the following modifications of the diffraction model,

$$\widetilde{S}_\ell = \frac{1}{1 + exp(-i\alpha)\, exp\left[(L-\ell)/\Delta\right]}\; \frac{\ell - L_o - i z\,(\ell)}{\ell - L_o - \frac{i}{2}\,\hat{\Gamma}(\ell)}\;. \qquad (5.5.5)$$

The first factor gives the usual diffraction model, the second, a Regge pole. It is advantageous to include the Regge term as a factor rather than as a summand since in this way the last factor in eq. (5.5.5) is unitary for $z = -\hat{\Gamma}/2$. The Regge pole lies at $\ell = L_o + \frac{i}{2}\,\hat{\Gamma}(\ell)$, the zero at $\ell = L_o + i z(\ell)$. For $\hat{\Gamma}$ and z , the following parametrizations were used: $\hat{\Gamma}(\ell) = \Gamma\left[1 + exp\{(\ell-L)/\Delta\}\right]^{-1}$ and $z = \left(\frac{1}{2}\Gamma - D\right)\left[1 + exp\{(\ell-L)/\Delta\}\right]^{-1}$ with $\Gamma > 0$ and $0 \leqslant D \leqslant \Gamma$. The resonance is purely elastic if pole and zero occur symmetrically about the real axis, $\Gamma = D$. It is purely inelastic for $D = 0$. The zero lies below the real axis for $D > \frac{1}{2}\Gamma$. In this way, the behaviour of the phase shifts for fixed E_{CM} is characterized, the Regge term being described by the two additional parameters Γ , D . These should be dependent on energy if $\widetilde{S}_\ell$ as a function of E_{CM} is to be described. The behaviour of $\eta_\ell = |\widetilde{S}_\ell|$ is shown on fig. 5.5.5 for $\Gamma = 7.02$, $D = 3.54$, $L_o = L = 18.1$, $\Delta = 1.62$ and $\alpha = 0.526$ (full dots and solid curve). The dip near $\ell = 18$ is caused by the Regge zero. Fig. 5.5.6 shows the angular distribution for nonidentical $^{16}O + ^{16}O$ scattering at 26 MeV generated by this choice of parameters. While for $\theta \lesssim 90^o$, the scattering is dominated by the diffraction pattern, except for interferences with the Regge pole, the backward angles display the typical Regge behaviour, characterized by oscillations with a period determined by $P_{\alpha_n(E_R)}(-\cos\theta)$.

In summary, we see that Regge poles are particularly suited to describe surface phenomena. Such phenomena are most important for intermediate absorption where an isolated Regge zero may dominate. They tend to be washed out for strong absorption.

5.5.3 Identification of Regge poles

Whether or not some oscillation at backward angles in a calculated angular

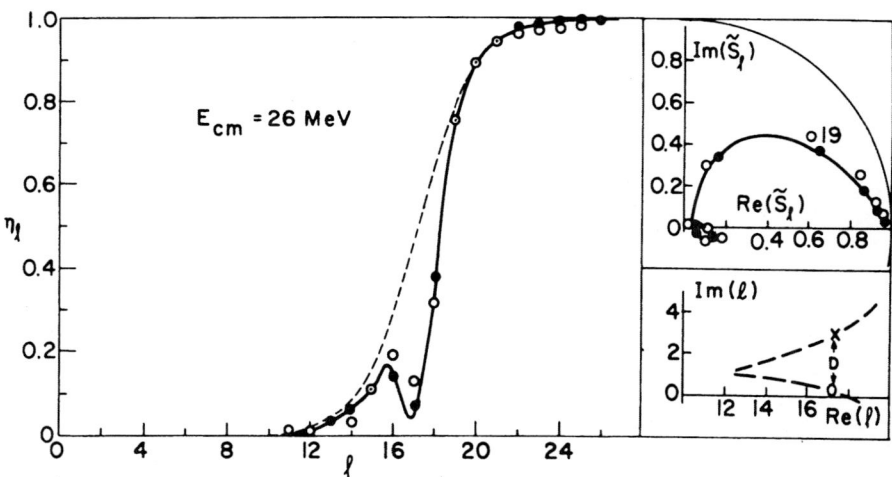

Fig. 5.5.5 The behaviour of $\eta_\ell = |\tilde{S}_\ell|$ versus ℓ for the
parameters given in the text. In the right part the
S-matrix elements and the Regge trajectories
for poles (x) and zeros (O) are shown. The open
circles are calculated from an optical-model
potential. (From [MC 71]).

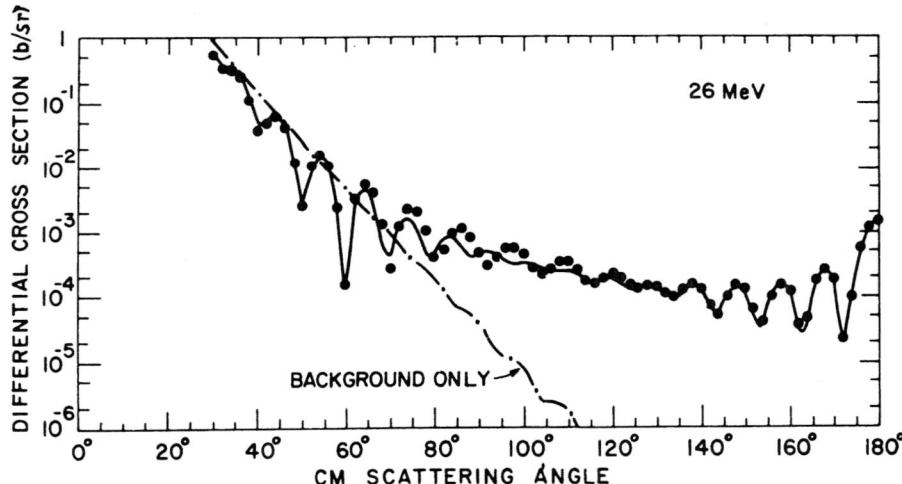

Fig. 5.5.6 The cross-section versus angle with Regge-pole dominance
for the parameter values given in the text. The points
are generated with an optical model. The dashed curve
is the result of the background term alone. (From [MC 71]).

distribution of charged particles is, indeed, caused by Regge behaviour can be checked as follows [LO 73] . One calculates the angular distribution twice, once for the actual problem which gives

$$k\,f(\theta) = k\,f_{C\ell b}(\theta) + \sum_{\ell=0}^{\infty} (2\ell+1)\exp(2i\sigma_\ell)\,\sin\delta_\ell\,\exp(i\delta_\ell)\,P_\ell(\cos\theta) , \qquad (5.5.6)$$

and the second time by artificially putting the Coulomb terms $f_{C\ell b}(\theta) = 0$ and $\sigma_\ell = 0$ in eq. (5.5.6). This is not the same as putting the charge of the projectile equal to zero which would also change the "nuclear" phase shifts δ_ℓ . If a Regge pole dominates the large angle scattering, the ratio of the two calculated cross-sections should be constant, since the oscillatory structure is the same in both cases. Using this test, one can show that Regge behaviour dominates large-angle scattering even in cases where a plot like the one shown in fig. 5.5.5 displays no dip. This will happen in cases when the Regge zero is still sufficiently far above the real axis, $\Gamma \gg D$.

5.5.4   Dynamical interpretation of Regge poles

The parametrizations presented so far leave open the question whether a Regge pole (or zero) is more than a convenient description of some features of elastic scattering. If it signifies some resonance process, two extreme and opposite inter- pretations are possible: (i) a Regge pole describes a family of resonances caused by mutual attraction of the two ions. In this case, the same family of resonances should be observed in inelastic channels. (ii) a Regge pole describes a surface effect in diffraction scattering just as in electromagnetic theory. There, scattering by a conducting sphere produces a surface wave which penetrates into the shadow region surrounding the forward hemisphere. This surface wave is essentially character- ized by the grazing angular momentum. It decays long before it surrounds the sphere and can thus be described as a Regge pole. This phenomenon occurs even without any attraction. Although the situation is not quite clear at this moment, indications exist that the truth lies somewhere between these two opposing views. These in- dications are now reviewed.

(i)   Regge poles as potential resonances [RI 72, SI 72] . An elastic resonance

should also show up in some strongly coupled inelastic channels. An identification
of the spin and parity is desirable. Singh et al. [SI 72] studied the reaction
$^{16}O(^{16}O, ^{12}C)^{20}Ne$ to three final states in ^{20}Ne. Narrow structures of about 300 keV
width and gross structures of about 1.5 MeV width were seen in the excitation
functions. These excitation functions show significant and strong angular correlations and cross correlations between different exit channels, especially for two
excited states in ^{20}Ne. The correlations observed experimentally are sufficiently
strong to rule out a purely statistical phenomenon. Moreover, the angular distributions seen rule out a diffraction pattern, and point to dominant contributions
with L = 16 for $E_{CM} \approx 23$ MeV, and with L = 18 for $E_{CM} \approx 27$ MeV, in accord with
the findings for elastic scattering of ^{16}O on ^{16}O . It is concluded that the gross
structure in the data is due to a resonance mechanism in the entrance channel
(doorway state). The fact that no correlations were observed in the reaction populating the ground state of ^{20}Ne is ascribed to the fact that the resonance(s) have
a small width for decay into this channel.

The best test in principle for the resonance hypothesis which seems not to
have been applied as yet, was suggested by Rinat [RI 72] and consists in an application of the Breit-Wigner formula for isolated resonances. It is assumed that a
single resonance dominates the cross-section (and herein lies the difficulty in
applying this test). It then follows that for two channels labelled 1 and 2 we
should find $\dfrac{d\sigma_{11}(E,\theta)}{d\Omega} \cdot \dfrac{d\sigma_{22}(E,\theta)}{d\Omega} = \left[\dfrac{d\sigma_{12}(E,\theta)}{d\Omega} \right]^2$. A relationship of this type
follows not only from the Breit-Wigner formula for isolated resonances, but also for
a doorway-state resonance.

(ii) This interpretation has recently been pursued by Fuller [FU 73] who has shown
that it leads to a very natural and simple description of heavy-ion induced transfer
reactions [FU 74, FU 75, MC 75] . The nucleus is pictured as being essentially
a black sphere, lit by an impinging plane wave as shown in fig. 5.5.7 . In the lit
region of the nuclear surface, a surface wave is formed which is essentially
characterized by the grazing angular momentum. This wave propagates as a creep wave
around the nucleus. In the shadow region, this leads to an illumination of the

nuclear surface. The creep waves decay exponentially and thus have a characteristic width in angle or correspondingly and according to the uncertainty relation, in angular momentum. Their behaviour is thus characterized by a complex angular momentum, the real part of which is essentially the grazing angular momentum, while the imaginary part is the width. In this way, one is naturally led to consider a Regge-pole description. The creep waves surrounding the nucleus interfere in the forward direction and there produce a characteristic Regge pattern. The relation of the wave function near the surface to the scattering amplitude has been studied by Austern [AU 75] . Neglecting Coulomb repulsion he finds that a cross-section localized near Θ is associated with a wave function localized at $\Theta + \frac{\pi}{2}$ near to the nuclear surface (cf. fig. 5.5.7) . If a Regge pole description applies, there exists an additional contribution from the angle $\Theta + \frac{3}{2}\pi$ on the nuclear surface (cf. fig. 5.5.7). It arises because the surface creep wave associated with a Regge pole illuminates the shadow region.

Technically, there exists a difference between the Regge description of the scattering amplitude as given in eq. (5.5.4), and the Regge description of the scattering wave function in the shadow region of the nuclear surface. Eq.(5.5.4) was obtained by replacing $(-)^{\ell} P_{\ell}(\cos\theta)$ by $P_{\ell}(-\cos\theta)$. It was remarked below eq. (5.5.3) that this procedure is most suitable at backward angles. For the description of the creep waves on the surface, it is important not to replace $(-)^{\ell} P_{\ell}(\cos\theta)$ by $P_{\ell}(-\cos\theta)$ as we are here interested in the forward direction.

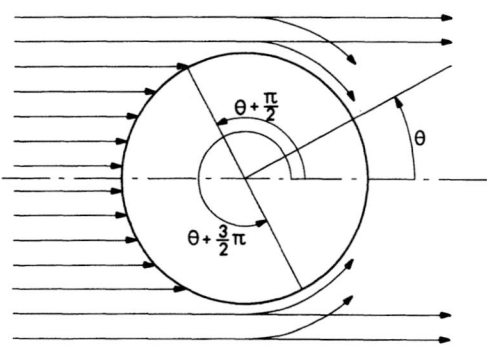

Fig. 5.5.7 The creep wave generated by the lit surface (schematic).

References

AL 65   V. De Alfaro and T. Regge, Potential Scattering (North-Holland, Amsterdam, 1965)

AR 73   A.G. Artukh, G.F. Gridnev, V.L. Mikheev, V.V. Volkov and J. Wilczynski,

Nucl.Phys. $\underline{A215}$ (1973) 91

AU 61   N. Austern, Ann. Phys. (N.Y.) $\underline{15}$ (1961) 299

AU 75   N. Austern, Phys. Rev. $\underline{C12}$ (1975) 128

BE 66 M.V. Berry,Proc.Phys.Soc.(London) $\underline{89}$ (1966) 479,

BE 71   M.C. Bertin, S.L. Tabor, B.A. Watson, Y. Eisen and G. Goldring, Nucl. Phys.

$\underline{A167}$ (1971) 216

BE 72   M.V.Berryand K.E.Mount,Rep.Progr.Phys.35 (1972) 315

BL 54   J.S. Blair, Phys. Rev. $\underline{95}$ (1954) 1218

BL 63   J.M. Blatt and V.F. Weisskopf, Theoretical Nuclear Physics (John Wiley & Sons,

New York - London, 1963)

BO 65   M. Born and E. Wolf, Principles of Optics (Pergamon, Oxford, 1965)

BR 72a R.A. Broglia, S. Landowne and A. Winther, Phys. Lett. $\underline{40B}$ (1972) 293

BR 72b R.A. Broglia and A. Winther, Phys.Lett. $\underline{4C}$ (1972) 153

BR 73   D.A. Bromley, Proc. of the International Conference on Nuclear Physics,

Munich, 1973 (North-Holland Publ.Comp., Amsterdam-London, 1973) p.38

BR 74   R.A. Broglia, S. Landowne, R.A. Malfliet, V. Rostokin and A. Winther,

Phys.Lett. $\underline{11C}$ $\underline{(1974)}$ 1

CO 70   A.A. Cowley and G. Heymann, Nucl. Phys. $\underline{146}$ (1970) 465

DO 75   C.D. Dover and J.P. Vary, in $\left[\text{HA 75}\right]$ p.1

EI 72   Y. Eisen and Z. Vager, Nucl.Phys. $\underline{A187}$ (1972) 219

EL 61   L.R.B. Elton, Nucl. Phys. $\underline{23}$ (1961) 681

FO 59   K.W. Ford and J.A. Wheeler, Ann. Phys. (N.Y.) $\underline{7}$ (1959) 259 and 287

FR 55   F.L. Friedman and V.F. Weisskopf, in "Niels Bohr and the Development of

Physics", ed. by W. Pauli (Pergamon Press, New York, 1955) p.134

FR 63   W.E. Frahn and R.H. Venter, Ann. Phys. (N.Y.) $\underline{24}$ (1963) 243

FR 65   N. Fröman and P. Fröman, JWKB Approximation (North-Holland, Amsterdam, 1965) p.86

FR 67   W.E. Frahn in "Fundamentals in Nuclear Theory", ed. by. A. de-Shalit and

C. Villi (IAEA, Vienna, 1967)

FR 72   W.E. Frahn, Ann. Phys. (N.Y.) $\underline{72}$ (1972) 524

FR 73   W.E. Frahn, in Extended Seminar on Nuclear Physics, Trieste 1973
(IAEA, Vienna, 1975) vol.I, p.157

FR 75   W.E. Frahn, in [HA 75] p.102

FU 73   R.C. Fuller, Nucl. Phys. $\underline{A216}$ (1973) 199; R.C. Fuller and Y. Avishai,
Nucl. Phys. $\underline{A222}$ (1974) 365

FU 74   R.C. Fuller and O. Dragun, Phys. Rev. Lett. $\underline{32}$ (1974) 617

FU 75   R.C. Fuller and K.W. McVoy, Phys. Lett. $\underline{55B}$ (1975) 121

GO 73   A. Gobbi, R. Wieland, L. Chua, D. Shapira and D.A. Bromley, Phys.Rev. $\underline{C7}$
(1973) 30

GR 60   K.R. Greider and A.E. Glassgold, Ann.Phys. (N.Y.) $\underline{10}$ (1960) 100

HA 74a H.L. Harney, P. Braun-Munzinger and C.K. Gelbke, Z.Physik $\underline{269}$ (1974) 339

HA 74b M.L. Halbert, C.B. Fulmer, S. Raman, M.J. Saltmarsh, A.H. Shell and
P.H. Stelson; Elastic Scattering of ^{16}O by ^{16}O, preprint (Oak Ridge Nat.
Lab., 1974)

HA 75   Classical and Quantum-Mechanical Aspects of Heavy-Ion Collisions,
ed. by. H.L. Harney et al. (Springer Verlag, Berlin - Heidelberg -
New York, 1975)

KN 74   J. Knoll and R. Schaeffer, Phys.Lett. $\underline{52B}$ (1974) 131

KN 75   J. Knoll and R. Schaeffer, preprint CEN-Saclay-DPh-T/75/40 (April 1975)
submitted to Ann. Phys. (N.Y.)

KO 75   T. Koeling and R.A. Malfliet, Phys.Lett. $\underline{22C}$ (1975) 181

LO 73   J.T. Londergan and K.W. McVoy, Nucl. Phys. $\underline{A201}$ (1973) 390

MA 69   J.V. Maher, M.W. Sachs, R.H. Siemssen, A. Weidinger and D.A. Bromley,
Phys. Rev. $\underline{188}$ (1969) 1665

MA 73   R.A. Malfliet, Symposium on Heavy Ion Transfer Reactions, Argonne, 1973
(Argonne, PHY-1973B) p.605; R.A. Malfliet, S. Landowne and V. Rostokin,
Phys.Lett. $\underline{44B}$ (1973) 238

MA 75   R.A. Malfliet, in [HA 75] p. 86

MC 60   J.A. McIntyre, K.H. Wang and L.C. Becker, Phys. Rev. $\underline{117}$ (1960) 1337

MC 70   M.R.C. McDowell and J.P. Coleman, Introduction to the Theory of Ion-Atom
Collisions (North-Holland, Amsterdam, 1970) chapter 2

MC 71   K.W. McVoy, Phys. Rev. $\underline{C3}$ (1971) 1104

MC 75   K.W. McVoy, in $\left[\text{HA } 75\right]$ p. 127

ME 61   A. Messiah, Quantum Mechanics (North-Holland, Amsterdam, 1961) vol. 1

ME 62   A. Messiah, Quantum Mechanics (North-Holland, Amsterdam, 1962) vol. 2

NE 66   R.G. Newton, Scattering Theory of Waves and Particles (McGraw-Hill, New York, 1966)

RA 66   G.H. Rawitscher, Nucl. Phys. $\underline{85}$ (1966) 259

RI 72   A.S. Rinat, Proc. of the Symp. on Four Nucleon Correlations and Alpha Rotator Structure, Marburg (Germany),1972, ed. by R. Stock (University of Marburg, 1972) p.245

SC 55   L.I. Schiff, Quantum Mechanics (McGraw-Hill Book Co., New York - Toronto - London, 1955)

SI 72   P.P. Singh, D.A. Sink, P. Schwandt, R.E. Malmin and R.H. Siemssen, Phys. Rev. Letts. $\underline{28}$ (1972) 1714

SI 73   R. da Silveira, Phys. Lett. $\underline{45B}$ (1973) 211

SI 74   R.H. Siemssen, in Nuclear Spectroscopy and Reactions, part B, p.233, ed. by J. Cerny (Academic Press, New York - London, 1974)

SO 59   A. Sommerfeld, Vorlesungen über Theoretische Physik, Optik (Geest & Portig, Leipzig, 1959)

TA 72   T. Tamura and H.H. Wolter, Phys. Rev. $\underline{C6}$ (1972) 1976

VE 63   R.H. Venter, Ann. Phys. (N.Y.) $\underline{25}$ (1963) 405

6.   Coulomb excitation

The process of Coulomb excitation is caused by the long-range electromagnetic interaction and takes place already at distances where the charge and mass distributions of the two heavy ions do not yet overlap. From a systematic point of view, it must therefore not be viewed in the spirit of the division of the S-matrix into a slowly and a rapidly varying part introduced in chapter 4, and it might thus have been more appropriate to place this chapter elsewhere in these lectures. That it has eventually found its place between the chapters on elastic and on inelastic and transfer collisions is due to the fact that some of the methods developed in the present chapter are also used in the description of inelastic processes caused by the nuclear interaction.

Already in the early stages of the study of nuclear reactions the possibility was discussed in the 1930's by Rutherford, Landau, Weisskopf and others of producing nuclear excitations by the well-known long-range electric field of the bombarding particles. Especially for incident energies so low that the Coulomb interaction prevents the colliding fragments from penetrating each other, such excitations can be studied without interference from the more complicated nuclear interactions. With almost exactly these words starts the classical review paper on Coulomb excitation by Alder et al. [AL 56] published in 1956. Much experimental and theoretical work has since been done, and information on nuclear structure, especially on transition rates and spin quantum numbers, has been obtained. The structure of rotational bands has been elucidated, and deformations have been studied. Reviews of this important field may be found in [AL 56, AL 66, BI 65, MC 74, AL 75]. Since the electric field produced by a heavy ion is proportional to Z , Coulomb excitation is a very important process in all of heavy-ion physics. We therefore review this topic, and emphasize the difference to and advantages over usual Coulomb excitation as well as the additional problems encountered in Coulomb excitation with heavy ions.

6.1   Qualitative considerations

Before proceeding, the reader should recall sections 2.1 and 2.2 and the relations given therein. Table 6.1.1 gives, for various fragments and energies, the numerical values of the Sommerfeld parameter η . We see that for energies below or at the Coulomb barrier, we usually have $\eta \gg 1$ in HI scattering. Looking at table 2.4.1 , we see that the classical approximation is valid for such energies up to

Table 6.1.1

Characteristic values for η in HI reactions

A_1, Z_1	A_2, Z_2	$E_{CM} = V_{CB}$ (MeV)	A_{red}	η
1, 1	200, 80	13	1	4
40, 20	200, 80	170	33	120
200, 80	200, 80	550	100	450

rather large scattering angles. Typical angular momenta for Coulomb trajectories have values around 10^3. Coulomb excitation is therefore usually described as follows. One calculates the time-dependent electromagnetic field produced at the location of the target by the projectile on its way along the classical trajectory.

The excitation process is due to the time-dependence of the electromagnetic field produced at the target. A typical characteristic frequency of this field has the value v_∞/a_c , where v_∞ is the relative velocity at infinite distance, and a_c the characteristic length of Coulomb scattering as defined by eq. (2.1.11b). In order that a state with excitation energy $\Delta E = \hbar\omega$ be excited, we require that $\omega \lesssim v_\infty/a_c$. It is, therefore, useful to define the "adiabaticity parameter" ξ by

$$\xi = \frac{\omega \, a_c}{v_\infty} = \eta \, \frac{\Delta E}{2 E_{CM}} \quad . \tag{6.1.1}$$

For $E_{CM} = V_{CB}$, ξ has the value

$$\xi = 0.05 \sqrt{A_{red}} \; \Delta E \left(\frac{R}{Z_1 Z_2} \right)^{\frac{3}{2}} \tag{6.1.2}$$

where ΔE is in MeV and R in fm . We expect a large excitation probability for $\xi \lesssim 1$. In the limit $\xi \gg 1$, the rate of change of the electromagnetic field with time becomes so small that the process approaches the adiabatic limit, hence the name for ξ . Because of the appearance of the factor $(Z_1 Z_2)^{3/2}$ in the denominator of eq. (6.1.2), ξ is smaller in HI reactions than for reactions induced by light particles, and we expect to populate fairly high-lying states not reached in ordinary Coulomb excitation. A typical example is the E1 giant dipole resonance with an excitation energy around 15 MeV in heavy nuclei.

The interaction energy is proportional to $Z_1 Z_2$. We thus expect the cross-section for Coulomb excitation to become large in HI reactions. In particular, first-order perturbation theory may not be adequate to take account of this interaction, and multiple excitation is the dominant process (see fig. 6.1.1).

For a freely propagating electromagnetic wave, the electric and magnetic field strengths are the same. This is not so in the vicinity of the particle for the electromagnetic field generated by a charged particle, for which the electric field strength is due to the charge, while the magnetic field strength is due to its current and thus a factor (v/c) smaller than the electric field. For fixed c.m.energy, HI are particularly slowly moving objects, in comparison with light projectiles.

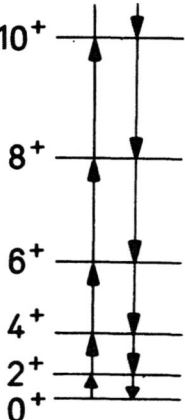

Fig. 6.1.1 Multiple excitation of a rotational band by E2 transitions

Therefore, electric transitions will be the dominant process in Coulomb excitation induced by HI. We thus do not consider magnetic processes in what follows.

For gamma radiation, the intensity of emission of an electric multipole field with multipolarity λ is characterized by the factor $(k_\gamma R)^{2\lambda+1} = (\omega R/c)^{2\lambda+1}$ where $\Delta E = \hbar\omega = \hbar k_\gamma c$, and where R is the nuclear radius. For Coulomb excitation, the ratio ω/c has to be replaced by a_c^{-1} , and the ratio $(a_c\omega/c)^{2\lambda+1} = \left(\frac{v_\infty \xi}{c}\right)^{2\lambda+1} \approx \left(\frac{\xi}{10}\right)^{2\lambda+1} \ll 1$ shows that in HI induced Coulomb excitation, higher multipoles figure much more prominently than in electromagnetic decays. As a consequence, one might hope to measure, e.g. hexadecapole moments through HI Coulomb excitation.

6.2 First-order perturbation theory

Although higher-order processes are important in Coulomb excitation with HI, it is instructive to first study the explicit formulas obtained in the frame of first-order perturbation theory. A detailed account may be found in [AL 56] . The physical picture used is that described in section 6.1 : The relative motion of the two fragments is described classically, and the time-dependent field generated by one fragment at the location of the other is calculated. The excitation due to this field is computed with the help of first-order perturbation theory. The cross-section for Coulomb excitation is then the product of two factors, the Rutherford cross-section and the probability $P_{i\to f}$ of excitation of a nuclear state f from the ground state i ,

$$\left(\frac{d\sigma}{d\Omega}\right)_{i\to f} = P_{i\to f}\left(\frac{d\sigma}{d\Omega}\right)_{Ruth} . \qquad (6.2.1)$$

Eq. (6.2.1) applies, of course, to the excitation of either fragment. Note that we have neglected the influence on the classical trajectory of the energy loss due to excitation. For unpolarized fragments, the probability $P_{i\to f}$ is given by

$$P_{i\to f} = \frac{1}{2I_i+1} \sum_{M_i M_f} |< f\, I_f M_f |\, b\, |\, i\, I_i M_i >|^2 \qquad (6.2.2)$$

where I_i, M_i, and I_f, M_f are spin and magnetic quantum number of the initial and final nuclear state, respectively. The amplitudes b are given by

$$\langle f I_f M_f | b | i I_i M_i \rangle = \frac{1}{i\hbar} \int\limits_{-\infty}^{+\infty} dt \, \langle f I_f M_f | \mathcal{H}(t) | i I_i M_i \rangle \, e^{i\omega_{if}t} \tag{6.2.3}$$

where $\mathcal{H}(t)$ is the interaction energy, and where $\hbar\omega_{if} = \Delta E$. Eq. (6.2.3) shows that the amplitudes b are essentially the matrix elements of the Fourier transform of $\mathcal{H}(t)$, taken at ω_{if}. This supports the qualitative consideration in section 6.1 which led to the introduction of the adiabaticity parameter ξ.

6.2.1 Electric excitation

For particle velocities small compared to c, the main interaction is the Coulomb energy,

$$\mathcal{H}_E(t) = \int \rho_n(\vec{r}) \, \varphi(\vec{r},t) \, d^3r \tag{6.2.4}$$

where $\rho_n(\vec{r})$ is the operator of the nuclear charge density, and $\varphi(\vec{r},t)$ is the Coulomb field produced by the projectile,

$$\varphi(\vec{r},t) = Z_1 e \left\{ \frac{1}{|\vec{r} - \vec{r}_p(t)|} - \frac{1}{r_p(t)} \right\} . \tag{6.2.5}$$

Here, $\vec{r}_p(t)$ is the vector connecting the c.m. of the two fragments. The term $1/r_p(t)$ has been subtracted in eq. (6.2.5) since it is already taken into account in calculating the classical trajectories. This operator acts on the c.m. coordinate of the nucleus. It cannot give rise to intrinsic excitation. To calculate the matrix elements in eq. (6.2.3), one expands $\varphi(\vec{r},t)$ into a multipole series and uses that the range of integration over $\vec{r}$ is restricted to the volume of the target, so that for $E_{CM} \lesssim V_{CB}$ we always have $r_p > r$. This gives

$$\varphi(\vec{r},t) = 4\pi Z_1 e \sum_{\lambda=1}^{\infty} \sum_{\mu=-\lambda}^{\lambda} \frac{1}{2\lambda+1} \frac{1}{r_p^{\lambda+1}} Y_{\lambda\mu}(\hat{r}_p) \, r^\lambda \, Y_{\lambda\mu}^*(\hat{r}) \tag{6.2.6}$$

where $\hat{r}_p$ and $\hat{r}$ are unit vectors in the direction of $\vec{r}_p$ and $\vec{r}$, respectively. The time dependence of $\mathcal{H}_E(t)$ is only contained in $\vec{r}_p(t)$ and thus does not affect the nuclear matrix elements. We define the matrix elements for an electric multipole transition by

$$\langle f I_f M_f | \mathfrak{M}(E\lambda,\mu) | i I_i M_i \rangle = \langle f I_f M_f | \int d^3r \, \rho_n(r) \, r^\lambda \, Y_{\lambda\mu}(\hat{r}) | i I_i M_i \rangle \qquad (6.2.7)$$

and find

$$\langle f I_f M_f | b | i I_i M_i \rangle = \frac{4\pi Z_1 e}{i\hbar} \sum_{\lambda\mu} \frac{1}{2\lambda+1} \langle f I_f M_f | \mathfrak{M}(E\lambda,\mu) | i I_i M_i \rangle \, S_{E\lambda,\mu} \qquad (6.2.8)$$

where we have introduced the orbital integrals

$$S_{E\lambda,\mu} = \int_{-\infty}^{+\infty} dt \, e^{i\omega_{if}t} \, Y_{\lambda\mu}(\hat{r}_p(t)) \left(\frac{1}{r_p(t)}\right)^{\lambda+1} . \qquad (6.2.9)$$

They are completely defined in terms of the classical trajectory $\vec{r}_p(t)$ and independent of nuclear structure. The nuclear structure information is contained in the electric multipole matrix elements which are the same as the ones for emission or absorption of gamma-radiation.

We define the reduced matrix elements by the convention

$$\langle i I_i M_i | \mathfrak{M}(E\lambda,\mu) | f I_f M_f \rangle = (-)^{I_i - M_i} \begin{pmatrix} I_i & \lambda & I_f \\ -M_i & \mu & M_f \end{pmatrix} \langle i I_i \| \mathfrak{M}(E\lambda) \| f I_f \rangle . \qquad (6.2.10)$$

Eq. (6.2.2) can be cast into the form

$$P_{i \to f} = \sum_\lambda P_{i \to f}(E\lambda) ,$$

$$P_{i \to f}(E\lambda) = \left(\frac{4\pi Z_1 e}{\hbar}\right)^2 \frac{B(E\lambda, I_i \to I_f)}{(2\lambda+1)^3} \sum_\mu |S_{E\lambda,\mu}|^2 ,$$

$$\qquad (6.2.11)$$

$$B(E\lambda, I_i \to I_f) = \frac{1}{2I_i+1} |\langle i I_i \| \mathfrak{M}(E\lambda) \| f I_f \rangle|^2 .$$

Because of the definitions (6.2.10) and (6.2.11), the "reduced transition probability" $B(E\lambda, I_i \to I_f)$ is related to the analogous quantity for γ-decay by

$$B(E\lambda, I_f \to I_i) = \frac{2I_i + 1}{2I_f + 1} \, B(E\lambda, I_i \to I_f) \ . \tag{6.2.12}$$

Eq. (6.2.11) can be simplified further by introducing special coordinates which are particularly suitable for the evaluation of the integrals $S'_{E\lambda,\mu}$. Without going into details, we write the final result for $(d\sigma/d\Omega)_{E\lambda}$, the contribution to the cross-section due to $E\lambda$ multipole absorption,

$$\left(\frac{d\sigma}{d\Omega}\right)_{E\lambda} = \left(\frac{Z_1 e}{\hbar \, v_\infty}\right)^2 a_c^{-2\lambda + 2} \, B(E\lambda, I_i \to I_f) \, f_{E\lambda}(\theta, \xi) \tag{6.2.13}$$

The dimensionless functions $f_{E\lambda}(\theta, \xi)$ contain all the information on the classical trajectory.

6.2.2   Discussion

The appearance of electric multipole operators in Coulomb excitation implies that the well-known selection rules for γ-radiation apply also in our case. These are

$$|I_i - I_f| \le \lambda \le I_i + I_f \qquad \text{(triangle inequality)} \qquad ,$$

$$\pi_i \, \pi_f = \left\{ \begin{array}{ll} (-)^\lambda & \text{for } E\lambda \\ (-)^{\lambda+1} & \text{for } M\lambda \end{array} \right\} \qquad \text{(parity selection rule)} \ . \tag{6.2.14}$$

We mention again that even in cases where a γ-decay has mixed multipolarity, E2 and M1 say, the Coulomb excitation proceeds mainly by E2 absorption. This is the reason that we have not written down the formulas for $M\lambda$ excitation. Formulas completely analogous to eq. (6.2.13) can, however, also be derived in the case of $M\lambda$ Coulomb excitation. In either case, the important dependence of the cross-sections on the energy, mass number and charge of projectile and target is contained in the functions $f_{E\lambda}(\theta, \xi)$, the analogous quantities $f_{M\lambda}(\theta, \xi)$ for magnetic transitions, and the functions $f_{E\lambda}(\xi) = \int d\Omega \, f_{E\lambda}(\theta, \xi)$ and $f_{M\lambda}(\xi) = \int d\Omega \, f_{M\lambda}(\theta, \xi)$ obtained by integration over the solid angle. These are now discussed. Fig. 6.2.1 shows $f(\xi)$ for various multipolarities as a function of ξ . This gives an idea about the relative contributions of the various multipoles for fixed ξ , and shows the essentially exponential fall-off of $f(\xi)$ with ξ for fixed multipolarity. The latter is due to increasing adiabaticity of the process. The functions $f_{E1}(\xi)$ and $f_{M1}(\xi)$ diverge logarithmically for $\xi \to 0$ while all other functions remain finite. This is due to the integration over angle and is caused by the long-range nature of the Coulomb field (remember that $\int d\Omega \, (d\sigma/d\Omega)_{Ruth}$ does not exist, either). The figure shows the preponderance of E1 excitation for small values of ξ , showing that the giant dipole E1 resonance can be strongly excited in HI

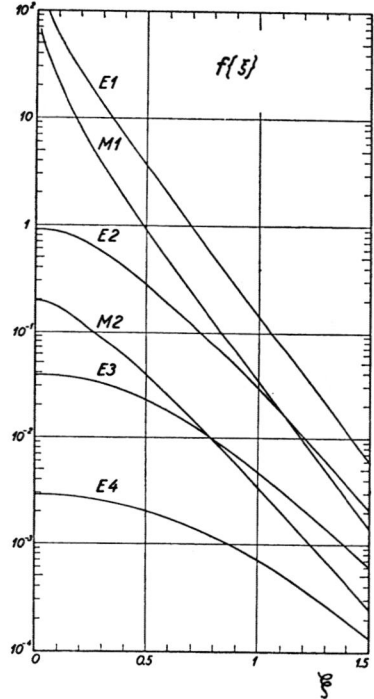

Fig. 6.2.1 The functions $f(\xi)$ for various multipolarities.
(From [AL 56]).

reactions. Fig. 6.2.2 shows the integrated cross-sections (in mb) for proton bombardment of Sn (Z = 50) versus proton energy for a hypothetical excitation energy of 200 keV and for various multipole transitions, where the reduced transition probability has been taken as 1 Weisskopf unit. Figs. 6.2.3 and 6.2.4 show the dependence of $f_{E1}(\theta, \xi)$ and $f_{E2}(\theta, \xi)$ on the scattering angle for various values of ξ . Observe the strong peaking in the forward direction which is characteristic for small ξ , and thus for HI reactions.

The nuclear states populated by Coulomb excitation can undergo γ-decay, and the angular distribution of the γ-rays can be measured. Without going into details, we just mention that experiments of this type as well as particle-gamma correlation data give the same (or even more) information as gamma-gamma correlation experiments.

The formulas given above were derived under the assumption that the target is excited, and that the projectile supplies the electric field. In HI reactions, this situation may, of course, be reversed. In the c.m. system, the formulas given above then still apply, provided that the quantities Z_1^2, ΔE, ξ and $B(E\lambda)$ are replaced by the corresponding values for the projectile .

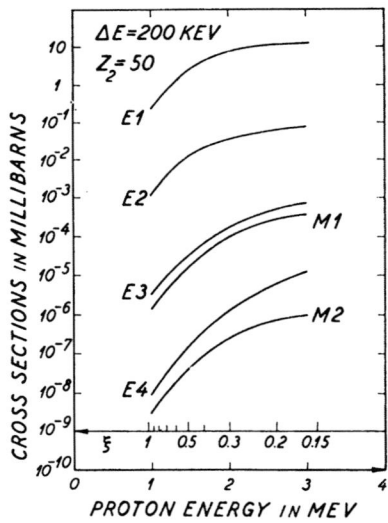

Fig. 6.2.2 Excitation cross-section for nuclear transitions of single-particle strength. (From [AL 56]).

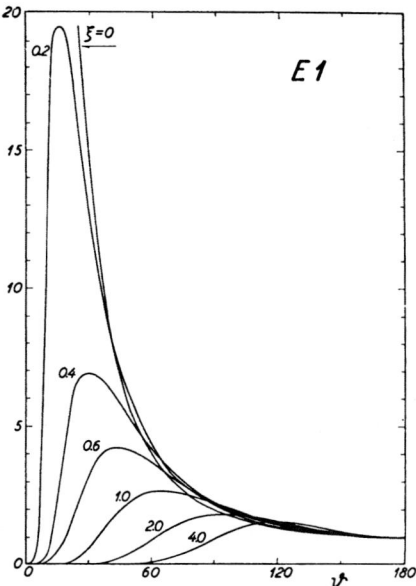

Fig. 6.2.3 Angular distribution of the inelastically scattered
            particles in classical approximation for E1
            excitations and for various values of the adiabaticity
            parameter ξ . (From [AL 56]).

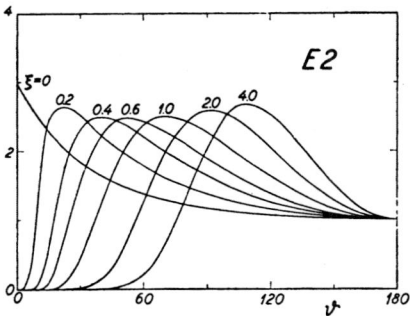

Fig. 6.2.4 The same as in fig. 6.2.3 for E2 transitions.
            (From [AL 56]).

6.2.3 Symmetrization of cross-sections

As mentioned above, the treatment presented in section 6.2.1 neglects the energy loss during the collision, and the effect this has on the classical trajectory. It is perhaps to be expected, however, that improved expressions for the excitation cross-sections are obtained by substituting for the particle velocity v_∞ entering into these expressions some appropriate mean value of initial and final velocity, v_i and v_f. Quantum mechanics suggests, in fact, that aside from a factor v_f/v_i, the cross-section should be symmetric in v_i and v_f. A straightforward way of symmetrizing the cross-sections consists in making the following substitutions, symbolised by arrows,

$$v_\infty^2 \longrightarrow v_i v_f ,$$
$$\alpha_c = \frac{z_1 z_2 e^2}{m v_\infty^2} \longrightarrow \frac{z_1 z_2 e^2}{m v_i v_f} ,$$
$$\xi = \frac{z_1 z_2 e^2}{\hbar} \frac{\Delta E}{2 E v_\infty} \longrightarrow \frac{z_1 z_2 e^2}{\hbar} \frac{|v_i - v_f|}{v_i v_f} . \tag{6.2.15}$$

The cross-sections take the form of eq. (6.2.13) with v_∞ replaced by v_i and with a_c and ξ given by the symmetrized expressions (6.2.15). A comparison with the quantum-mechanical results shows that the symmetrized expressions give a much better approximation than the formula (6.2.13) without symmetrization. This approximation is particularly good for large η and small ξ . i.e. for HI reactions. This is perhaps to be expected since the classical theory works best for HI .

6.2.4 Survey of the approximations

Some of the more important approximations introduced above are now reviewed. Most important among these are the classical approximation for the trajectories, and the use of first-order perturbation theory. These points are discussed in some detail in sections 6.3 and 6.4 . In restricting ourselves to energies below the Coulomb barrier, $E_{CM} < V_{CB}$, we have also neglected the nuclear interaction between projectile and target. Experience shows that this approximation is somewhat critical and holds only if the distance r_{min} of closest approach is considerably larger than the sum R of the nuclear radii. If one wishes to observe higher-order effects with-

out nuclear interference, one must require $r_{min} \geq R + 4$ fm . In writing down the
interaction, we have neglected the retardation, i.e. relativistic effects, and we
have neglected the fields produced by the quadrupole, hexadecapole etc. moments of
the projectile. While no estimate seems available for these fields, the relativistic
corrections are of the order v_∞^2/c^2 and are thus quite small. Atomic effects also
have to be considered. We have neglected the screening of the Coulomb field due to
the presence of electrons. So long as r_{min} is considerably smaller than the Bohr
radius of a K-shell electron, this is a good approximation. It was pointed out by
Greiner, however, that K-shell electrons orbiting the two colliding HI jointly may
produce an additional binding energy of ≈ 2 MeV and thus a sizeable modification of
the Coulomb potential, and of the classical trajectory. Finally, we have considered
the intrinsic wave function of projectile and target in the c.m. frame for either
particle, although the theory was developed in the c.m. frame of the entire system
(projectile plus target). It is, therefore, necessary to transform the wave functions
into this latter system. One usually neglects the acceleration and performs a
Galileo transformation, which corresponds to a unitary transformation of the wave
functions. As a consequence, these functions become multiplied by a time-dependent
phase. This phase can usually be neglected for Coulomb excitation, although it is of
some importance in transfer reactions when interference patterns are relevant.

6.3   Higher-order effects

The probabilities of Coulomb excitation calculated in the frame of first-
order perturbation theory in section 6.2 are usually quite small. For instance,
α-particles with $E_{CM} = 6$ MeV striking a target with $Z_2 = 70$ and a B(E2) =
$= 5 \cdot e^2 \cdot 10^{-48}$ cm^4 (which is already quite large) give, for $\xi \ll 1$, a transition
probability $P_{i \to f} \simeq 0.01$. However, the transition probability increases with
increasing E_{CM} , typical for HI collisions, essentially as E_{CM}^3 . Moreover, the
direct transition to the final state may be inhibited. In either case, higher-order
processes become important. In this section, we study such processes but keep the
classical approach, that is we describe the relative motion of the two fragments
in terms of the classical trajectory. We only take into account the electromagnetic
field produced by one fragment at the location of the other in higher orders than

the first. References are $\begin{bmatrix} AL\ 56,\ BO\ 68,\ PE\ 72, MC\ 74,\ AL\ 75 \end{bmatrix}$.

6.3.1 The coupled equations

We accordingly expand the wave function $|\psi(t)>$ of the target in terms of the eigenstates $|\psi_r>$ of the nuclear Hamiltonian H_o , with corresponding eigenvalues ε_r ,

$$|\psi(t)> = \sum_{r=0}^{\infty} c_r(t)\ exp\left(- i\varepsilon_r t/\hbar\right)|\psi_r> . \tag{6.3.1}$$

The expansion coefficients obey the condition $c_r(-\infty) = \delta_{ro}$ since the target is originally in its ground state. The occupation probability of the state r after the collision is given by $P_{o \to r} = |c_r(+\infty)|^2$, the cross-sections again have the form of eq. (6.2.1). The time-dependent coupled equations for the expansion coefficients $c_r(t)$ can now simply be derived from the time-dependent Schrödinger equation for $|\psi(t)>$ with the Hamiltonian $H_o + \mathcal{H}(t)$. Using the orthonormality of the $|\psi_r>$, one finds

$$i\hbar \dot{c}_s(t) = \sum_{r=0}^{\infty} <\psi_s|\mathcal{H}(t)|\psi_r>\ e^{i\omega_{sr}t}\ c_r(t) \tag{6.3.2}$$

where $\hbar\omega_{sr} = \varepsilon_s - \varepsilon_r$ and $c_r(-\infty) = \delta_{ro}$.

These equations can be solved numerically if the sum over r is truncated, and the nuclear matrix elements $<\psi_s|\mathcal{H}(t)|\psi_r>$ are evaluated as in section 6.2 . A program for this solution exists $\begin{bmatrix} WI\ 65 \end{bmatrix}$. To understand better the nature of the higher-order contributions, we solve eqs. (6.3.2) by iteration with respect to $\mathcal{H}(t)$. Up to terms of second order this yields

$$c_f^{(1)}(t) = \frac{1}{i\hbar} \int_{-\infty}^{t} <\psi_f|\mathcal{H}(t)|\psi_o>\ e^{i\omega_{fo}t} ,$$

$$c_f^{(2)}(t) = \frac{1}{i\hbar} \sum_m \int_{-\infty}^{t} <\psi_f|\mathcal{H}(t)|\psi_m>\ e^{i\omega_{fm}t}\ c_m^{(1)}(t) . \tag{6.3.3}$$

The transition probability $P_{o \to f}$ has the form (we write out only terms up to third order in $\mathcal{H}(t)$)

$$P_{0 \to f} = |c_f^{(1)}(\infty)|^2 + 2 \, \mathcal{Re} \left\{ c_f^{(1)}(\infty) \, c_f^{(2)*}(\infty) \right\} + \dots \quad . \tag{6.3.4}$$

We see that $1^{st}-$ and 2nd-order processes interfere. This is the reason for the importance of 2nd-order processes even in cases where they are fairly weak. Without working out eq. (6.3.4) in full detail, we now discuss some situations where $2^{nd}-$ order processes are important.

6.3.2 The reorientation effect

An interesting case where the interference between first and second-order terms in eq. (6.3.4) is significant is that where the intermediate state $|\psi_m\rangle$ differs from the final state $|\psi_f\rangle$ only in the magnetic quantum number. This applies mostly to E2 absorption by even-even nuclei. Then, $|\psi_m\rangle$ and $|\psi_f\rangle$ are different magnetic substates of the excited 2^+ state . Hence the name reorientation effect for this phenomenon . The matrix element $\langle \psi_f | \mathcal{H}(t) | \psi_m \rangle$ is proportional to the quadrupole moment Q_2 of the excited 2^+ state, and the interference term in eq. (6.3.4) is, therefore, __linear__ in Q_2. A precise determination of $P_{0 \to f}$ thus gives the possibility of determining magnitude __and sign__ of Q_2. For $E_{CM} \gg \Delta E$, the interference term can be evaluated approximately. Without going into details, we just quote the result $\left[\text{BO } 68 \right]$. We define the quantity r as the ratio of $2 \, \mathcal{Re} \left\{ c_f^{(1)}(\infty) \, c_f^{(2)*}(\infty) \right\}$ over $|c_f^{(1)}(\infty)|^2$, i.e. as the relative contribution of the reorientation effect. One finds, for $E_{CM} \gg \Delta E$,

$$r \approx \frac{A_{red}}{Z_2} \, \Delta E \, \langle f I_f \| \mathcal{M}(E2) \| f I_f \rangle \, K(\xi, \theta) \quad . \tag{6.3.5}$$

Note that the only dependence of this expression on energy and angle resides in the function $K(\xi, \theta)$ which is shown for some cases in fig. 6.3.1 . The largest reorientation effects are obtained in the backward direction where K approaches unity. Table 6.3.1, taken from $\left[\text{PE } 72 \right]$, shows typical r-values in the backward direction for ^{152}Sm as target with a quadrupole moment $Q_2 = 2$ barns. We see that large effects are to be expected for HI. For such large values of Q_2, other higher-order processes

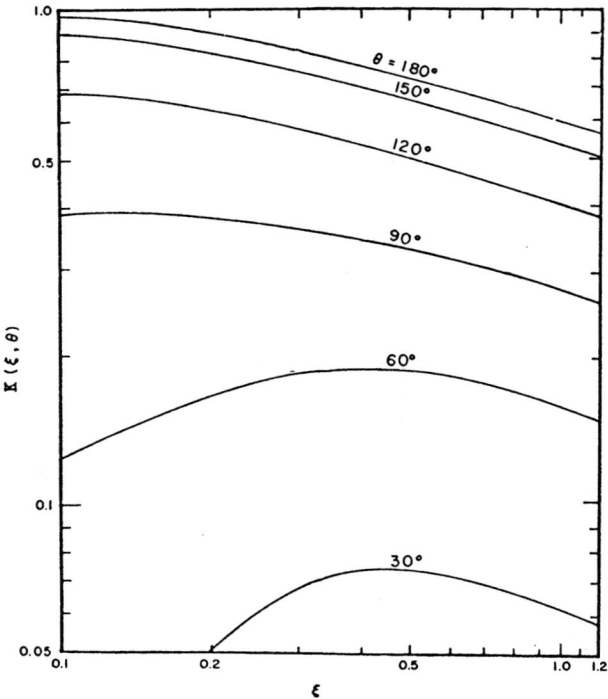

Fig. 6.3.1 The function K(ξ,θ) for some values of θ as a
function of ξ. (From [BO 68]).

Table 6.3.1 Values of r for various projectiles with mass A_1
bombarding ^{152}Sm. (From [PE 72]).

A_1	16	32	79	120
r	7%	12%	24%	32%

also become important. However, one can use the fact that for small ζ , and $\theta \gtrsim 90°$, $K(\zeta,\theta)$ depends very little on energy (see fig. 6.3.1) to separate the reorientation effect. To analyze the data, a qualitative discussion like the one given here does, of course, not suffice. Higher-order effects must usually be taken into account by a numerical solution of the coupled equations (6.3.2) .

6.3.3   Double E2 excitation

Another important second-order effect is that of double E2-excitation. Starting from a 0^+ ground state, it leads via a 2^+ state to the 4^+ state and interferes with the first-order E4-excitation populating the 4^+ state directly from the ground state. Since the direct E4-excitation probability is small, the second-order process is important. Biggest E4 contributions are found for small values of Z_1 and A_1 where the direct E4-excitation contributes about 5% to the overall probability. For such small values of Z_1 and A_1 (α-particles are often used, and $\eta \approx 10$) quantum-mechanical corrections are not negligible and complicate the analysis of the data, as do other higher-order effects. Fig. 6.3.2 shows the measured hexadecapole moments for several actinides. By applying the rotator model, one can deduce from these data

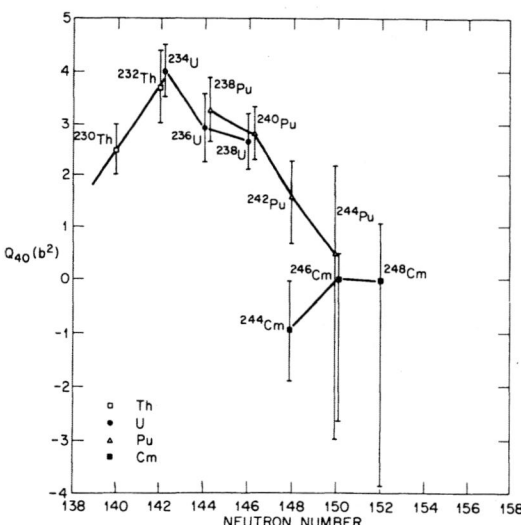

Fig. 6.3.2   Measured hexadecapole moments for several actinides.
(From [MC 72]).

the hexadecapole deformation of the ground state. Combining this information with the one for quadrupole deformation for the ground state of ^{234}U, one finds the shape shown in fig. 6.3.3 .

6.3.4   Electric dipole polarization

Another second-order process is the virtual excitation of the giant dipole resonance from the ground state, followed by deexcitation back to the ground state. The contribution of this process to the elastic scattering can be simply calculated in the adiabatic limit, $\xi \gg 1$. In this limit, one simply calculates the electro-static dipole moment induced in the target by the field of the projectile. The calculation can be carried out using standard, time-dependent perturbation theory. The resulting potential modifies the pure Coulomb trajectory. Since the strength of the polarizing field is proportional to the charge of the projectile, the effect is largest for heavy ions (for fixed E_{CM}) .

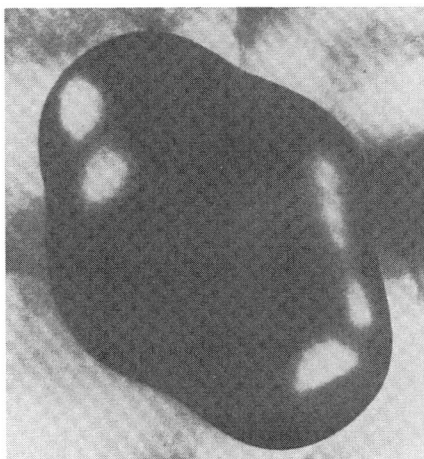

Fig. 6.3.3   Shape of the ^{234}U nucleus as deduced from E2 and E4 transition moments in the liquid drop model. (From [MC 72]).

6.3.5   Illustrative picture of the mechanism of multiple Coulomb excitation

(From [MC 74])

We consider Coulomb excitation of ^{238}U, bombarded with ^{40}Ar ions with 140 MeV (which corresponds to E_{CM} = 120 MeV, while $V_{CB} \approx$ 180 MeV)   scattered through an angle of 160 degrees. This situation is shown in fig. 6.3.4 . The characteristic parameters are a_c = 10 fm and η = 140 , which shows that the classical approximation is good. In fig. 6.3.4 we see the electrostatic forces acting on the deformed ^{239}U nucleus, decomposed into components tangential and normal to the surface. The tangential components produce a torque that puts the ^{238}U nucleus into rotating motion. The quantum-mechanical analogue of this rotation is excitation of ^{238}U, due to the quantization of the angular momentum. The rotational spectrum of ^{238}U is shown on the right-hand side of fig. 6.3.5 .

This simple picture brings into evidence several features of Coulomb excitation. (i) The torque on the target, and hence the excitation probability, increases with the charge of the projectile. (ii) The torque increases with decreasing distance of closest approach. (iii) The torque increases with deformation, i.e. with decreasing excitation energy of the rotational levels. The quantification of this picture makes it possible to obtain information about nuclear shapes from Coulomb excitation.

Along the hyperbolic path of fig. 6.3.4 several dots are shown which indicate the position of the ^{40}Ar projectile at constant time intervals of $2 \cdot 10^{-22}$ sec. When the ^{40}Ar projectile has the position indicated in the figure, the torque has only 1/5 of its maximum value. Most of the torque will thus be exerted during a time interval of about $3 \cdot 10^{-21}$ sec. This is to be compared with the time interval used in estimating the adiabaticity parameter $\xi = \alpha_c \omega / v_\infty$. Writing $\omega = 2\pi/\tau_{rot}$ where the rotation time $\tau_{rot} = 9.2 \cdot 10^{-20}$ sec for an excitation energy of 45 keV , and putting $\xi = \dfrac{\tau_{coll}}{\tau_{rot}}$, we find $\tau_{coll} = 2\pi a_c / v_\infty$ which yields $\tau_{coll} = 2.3 \cdot 10^{-21}$ sec in our example, in good agreement with the number given above. We also see that $\xi \approx 0.03$. $\xi \ll 1$ or $\tau_{coll} \ll \tau_{rot}$ means that the frequency spectrum contained in the time-dependence of the torque embraces the frequencies for single- and for multiple excitation. In this situation, multiple excitation is important, and the excitation

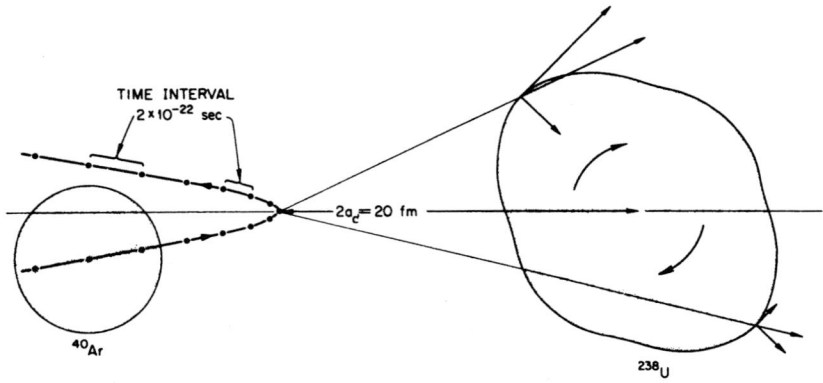

Fig. 6.3.4 Scattering of ^{40}Ar by ^{238}U. (From [MC 74]).

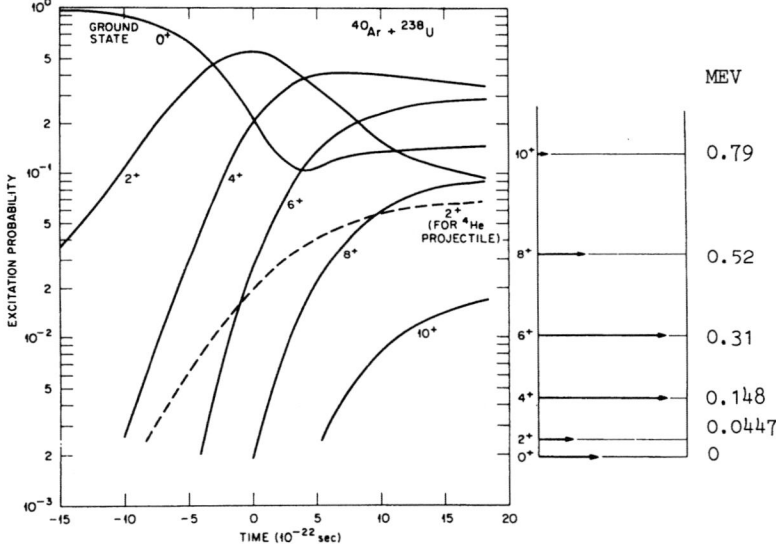

Fig. 6.3.5 Excitation probabilities of the rotational states
of ^{238}U as function of time. (From [MC 74]).

probabilities of the various nuclear states must be worked out by numerical integration of the eqs. (6.3.2) which use the nuclear matrix elements as input. For the
example discussed above, the time-dependence of the occupation probability of various
levels is shown on the left-hand side of fig. 6.3.5 . Time $t = 0$ denotes the
distance of closest approach. The importance of higher-order effects is clearly seen
in the figure. Eventually, the largest occupation probabilities are those for the
4^+ and 6^+ levels, while the ground state is fairly depleted.

6.4 Quantum-mechanical corrections

It is important to investigate the accuracy of the semiclassical theory.
This is particularly necessary in view of the smallness of second-order effects
such as the reorientation effect which is used to extract valuable information from
the data. This was done in $\begin{bmatrix} SM\ 68,\ AL\ 69,\ AL\ 72a,\ AL\ 72b \end{bmatrix}$. In first-order perturbation theory, the transition amplitude $b_{i \to f}^{(\lambda \mu)}$ for a multipolarity λ with
magnetic quantum number μ can be parametrized in the form

$$b_{i \to f}^{(\lambda \mu)} = \left(b_{i \to f}^{(\lambda \mu)} \right)_{sc} r_{\lambda \mu} \exp \left(i \psi_{\lambda \mu} \right) \tag{6.4.1}$$

where $r_{\lambda \mu}$ and $\psi_{\lambda \mu}$ are real, and where $\left(b_{i \to f}^{(\lambda \mu)} \right)_{sc}$ is the expression obtained from
the semiclassical theory reviewed in section 6.2 . The correction $r_{\lambda \mu}$ to the abso-
lute value differs from unity by terms of order η^{-2} and can, therefore, safely
be replaced by unity. The correction $\psi_{\lambda \mu}$ behaves as η^{-1} where η is the
symmetrised quantity introduced in subsection 6.2.3 . We see that the semiclassical
theory is expected to yield good values for cross-sections in first order, while for
interference effects the quantum-mechanical corrections become important and can be
neglected only for very large values of η $\left(\eta \geq 100 \right)$.

6.4.1 The coupled equations

The Schrödinger equation for the scattering process is transformed into a
set of coupled differential equations. This transformation is described in section
7.3.1. The following assumptions are used.

(i) The projectile has no internal degrees of freedom, and spin zero. It
acts like a point charge.

(ii) $E_{CM} < V_{CB}$ so that the fragments do not mutually penetrate, and only electromagnetic effects need be considered.

For a given total spin J of the system, the radial wave functions $g_{\ell I J}(r)$ of the relative motion of the two fragments, labelled by the relative orbital angular momentum ℓ and by the spin I of the state of the projectile in addition to J, obey the equations

$$\left\{ \frac{d^2}{dr^2} + k_I^2 - \frac{2\eta_I k_I}{r} - \frac{\ell(\ell+1)}{r^2} \right\} g_{\ell I J}(r) = \sum_{I'\ell'} V_{\ell I, \ell' I'}^J(r) \, g_{\ell' I' J}(r) \tag{6.4.2}$$

where $\hbar k_I$ is the relative momentum corresponding to an excitation of state I, and where η_I is similarly defined. The coupling matrix elements $V_{\ell I, \ell' I'}^J(r)$ have the form

$$V_{\ell I, \ell' I'}^J(r) = \sqrt{16\pi} \;\; \frac{m \, Z_1 e}{\hbar^2} \; .$$
$$\cdot \sum_\lambda (-)^{J+I'+\lambda} \begin{pmatrix} \ell \, \ell' \, \lambda \\ 0 \, 0 \, 0 \end{pmatrix} \left\{ \begin{matrix} J \, \ell' \, I' \\ \lambda \, I \, \ell \end{matrix} \right\} \sqrt{\frac{(2\ell+1)(2\ell'+1)}{2\lambda+1}} \;\; \frac{\langle I \| \mathcal{M}(E\lambda) \| I' \rangle}{r^{\lambda+1}} \; . \tag{6.4.3}$$

As in the preceding sections, we have only considered the electric field. Aside from the angular momentum coupling coefficients and the overall normalization, the form of eq. (6.4.3) is intuitively expected. The eqs. (6.4.2) are solved numerically with the boundary condition that an incoming wave exists only in the channel ℓ, I and the elements $S_{\ell I, \ell' I'}^J$ of the S-matrix are obtained from the asymptotic form of the solutions. The S-matrix then yields the cross-section. This procedure is extremely tedious, for two reasons. The long range of the Coulomb field and of the coupling term makes it necessary to integrate the equations out to about 200 fm. For heavy ions, partial waves up to $\ell = 1000$ have to be taken into account. It is, therefore, of interest to find good approximations to these equations. This is possible by using the JWKB method described in section 5.1 . As shown in [AL 69] , the use of the JWKB approximation results in the (symmetrized) coupled equations of the

semiclassical theory which thus contain most of the quantal corrections. More precisely, the approximations used are: (i) suppression of contributions from distances
less than the classical turning point, (ii) use of the JWKB approximation for
distances larger than the classical turning point, (iii) restriction to large angular momenta, $\ell \gg 1$.

The main difference between the semiclassical and the quantal cross-section
thus arises not from the solution of the coupled equations which determines the
transition amplitudes, but rather from the way these amplitudes are connected with
the cross-sections. Just as in the diffraction theory of elastic scattering, one
arrives at the semiclassical expression for the cross-section if one replaces the
sum over ℓ-values by an integral and uses the method of steepest descent (or of
stationary phase). It is at this point that a major error is committed.

6.4.2 Discussion of the quantum-mechanical corrections

We first discuss the corrections to first-order transitions. For this purpose
we write the cross-section in the form

$$\left(\frac{d\sigma}{d\Omega}\right)_{E\lambda} = \left(\frac{d\sigma}{d\Omega}\right)_{E\lambda}^{sc} R(\theta,\eta,\xi) \qquad (6.4.4)$$

where $(d\sigma/d\Omega)_{E\lambda}^{sc}$ is the semiclassical expression (6.2.13) symmetrized according to
(6.2.15), and where $R(\theta,\eta,\xi)$ is the correction factor. An analogous quantity $R(\eta,\xi)$
can be defined as the correction factor for the integrated (symmetrized) semiclassical
cross-section. Note that $R(\eta,\xi)$ differs from $\int d\Omega\, R(\theta,\eta,\xi)$. For various values
of ξ and for E1 and E2 absorption, the function $R(\eta,\xi)$ is plotted versus η^{-1}
in fig. 6.4.1 . We emphasize that the correction would become very much smaller (and,
in fact, negligible for $\eta \gtrsim 2$) if the semiclassical cross-section were to be replaced by the JWKB approximation for the scattering amplitudes and the quantum-
mechanical cross-section formula. We see that in any case, the corrections for
E1 absorption are quite small. Those for E2 absorption follow the $1/\eta^2$ pattern
discussed above.

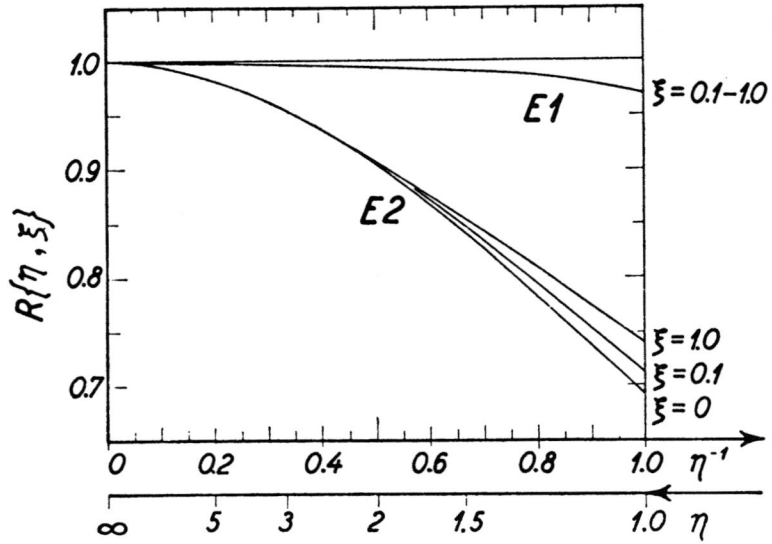

Fig. 6.4.1 The function $R(\eta, \xi)$ versus η^{-1}. (From [AL 56])

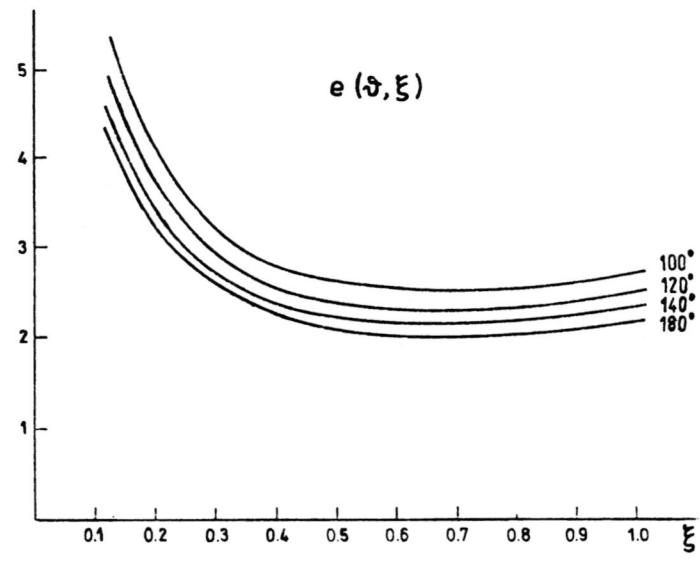

Fig. 6.4.2 The function $e(\theta, \xi)$ versus ξ. (From [AL 72a])

To discuss the quantal corrections to the reorientation effect, we write the quantum-mechanical differential cross-section in the form

$$\left(\frac{d\sigma}{d\Omega}\right)_{2+} = \left(\frac{d\sigma}{d\Omega}\right)^{(1)}_{2+} \left\{ 1 + r(\theta,\xi) + \frac{1}{\eta} \chi^{(2)}_{2\to 2} \, e(\theta,\xi) \right\} \tag{6.4.5}$$

The upper index (1) denotes the first-order-term, r is the semiclassical quantity introduced in eq. (6.3.5), $\chi^{(2)}_{2\to 2}$ is a strength parameter proportional to the quadrupole moment, and $e(\theta,\xi)$ contains the quantum-mechanical corrections. The last term in eq. (6.4.5) shows that these are of order $1/\eta$, as we expect. The function $e(\theta,\xi)$ is plotted versus ξ for various angles in fig. 6.4.2 . We see that $e(\theta,\xi)$ becomes large for small ξ . For heavy ion reactions, where η is large, the quantal corrections to the reorientation effect are negligible so long as $\xi \gtrsim 0.3$. We finally mention that quantal corrections for n^{th} order processes behave like n/η .

6.5 Methods of detecting Coulomb excitation [MC 74]

Two methods are extensively used: One either measures the energy spectrum of the scattered projectile, or the radiation (gamma rays or internal conversion electrons) following the excitation. We now briefly discuss the drawbacks and advantages of either method.

6.5.1 Direct measurement of the spectrum of the scattered particles

This method has the advantage that the absolute cross-sections can be measured with good accuracy (of about 1%) since the total cross-section equals the Rutherford cross-section. Many standard problems like target thickness, particle flux, etc. can thus be eliminated by a measurement of the ratio of peak areas. The method suffers from small counting rates of inelastic versus elastic events, from limited energy resolution, and because of the nuclear information which can be extracted from this type of data is rather limited. To improve on the energy resolution, a beam with good energy homogeneity, thin targets, and a high-resolution spectrometer is needed. For instance, to measure Coulomb excitation of ^{114}Cd with ^{84}Kr projectiles, of about 250 MeV, optimal conditions still lead to an overall energy spread of about 400 keV, whereas the first excited 2^+ state in ^{114}Cd has an excitation energy of only 558 keV . For rotational states, it is hopeless to use this method with beams of heavy ions.

6.5.2   Detection of deexcitation gamma rays

The advantage of this method - excellent energy resolution and good counting rates - are essentially preserved as one switches to heavier projectiles. The energy resolution suffers from the Doppler shift of the emitted gammas. The line broadening can be much larger than the resolution of the Ge(Li) detector. On the other hand, this shift can in itself be used to obtain valuable information about lifetimes in the psec region.

References

AL 56   K. Alder, A. Bohr, T. Huus, B. Mottelson and A. Winther, Rev.Mod.Phys. $\underline{28}$ (1956) 432

AL 66   K. Alder and A. Winther, Coulomb Excitation (Academic Press, New York, 1966)

AL 69   K. Alder and K.A. Pauli, Nucl. Phys. $\underline{A128}$ (1969) 193

AL 72a K. Alder, Proc. of the Heavy Ion Summer Study, Oak Ridge, 1972 (Oak Ridge Nat Lab. Report CONF-720669) p.94

AL 72b K. Alder, F. Roesel and R. Morf, Nucl. Phys. $\underline{A186}$ (1972) 449

AL 75   K. Alder and A. Winther, Electromagnetic Excitation - Theory of Coulomb Excitation with Heavy Ions (North-Holland Publ. Comp., Amsterdam - Oxford, 1975)

BI 65   L.C. Biedenharn and P.J. Brussaard, Coulomb Excitation (Clarendon Press, Oxford, 1965)

BO 68   J. de Boer and J. Eichler, Adv. Nucl. Phys. $\underline{1}$ (1968) 1

MC 72   F.K. McGowan, Proc. of the Heavy Ion Summer Study, Oak Ridge, 1972 (Oak Ridge Nat. Lab., Report CONF-720669) p.38

MC 74   F.K. McGowan and P.H. Stelson, in Nuclear Spectroscopy and Reactions, part C, edited by J. Cerny (Academic Press, New York, London, 1974) p.3

PE 72   D. Pelte, Überlegungen zu einer Nachbeschleunigungsstrecke für mittelschwere Ionen am Heidelberger MP-Tandem-Beschleuniger (MPI Heidelberg, MPI-H-1972-V26) p.21

SM 68   U. Smilansky, Nucl. Phys. $\underline{A112}$ (1968) 185

WI 65   A. Winther and J. de Boer, A Computer Program for Multiple Coulomb Excitation, Progress Reports (State University and California Institute of Technology, Rutgers, 1965)

7.   Inelastic scattering and transfer reactions

As mentioned in chapter 3 (see fig. 3.4.1), there are three classes of heavy-ion collisions: distant collisions, close collisions, and grazing collisions. We are here concerned with the latter. At bombarding energies only slightly higher than the Coulomb barrier, the kinetic energy in the interaction region is small, and the two particles can fuse easily if they reach the interaction region. At higher energies, the situation changes, and there exists an intermediate range of impact parameters where the two particles interact without fusing, since this is prohibited by the high kinetic energy and the high centrifugal barrier (see fig. 7.1 which displays the situation depicted in fig. 3.4.4 in a different fashion). In addition to Coulomb excitation, the nuclei may also get excited through the nuclear forces in such a grazing collision, or transfer of one or several nucleons may take place. The description of these processes, which may partly interfere with Coulomb excitation, forms the content of the present chapter. We focus attention on the <u>dynamics</u> of these reaction processes. A discussion of nuclear structure information such as spectroscopic factors contained in the data would exceed the frame of these lectures. We exclude also deeply inelastic reactions (cf. chapter 2) which dominate the cross-section in collisions between very heavy nuclei.

We first describe some general properties of transfer reactions which can be understood classically ([SI 71, BR 72a] , section 7.1) or semiclassically ([BR 72b] , section 7.2). We introduce the quantum-mechanical coupled-channels formalism (section 7.3) which contains the distorted-wave Born approximation and the description of the so-called elastic transfer processes as special cases. Finally, we discuss the parametrization of phase shifts (smooth cut-off model, Regge-pole model) in inelastic scattering and transfer reactions (section 7.4).

7.1   Classical considerations   [BR 72a]

Classical considerations were applied to transfer processes by Siemens et al. [SI 71] . We follow the more detailed subsequent paper by Brink [BR72a] . We consider the transfer of a single nucleon n in the collision process

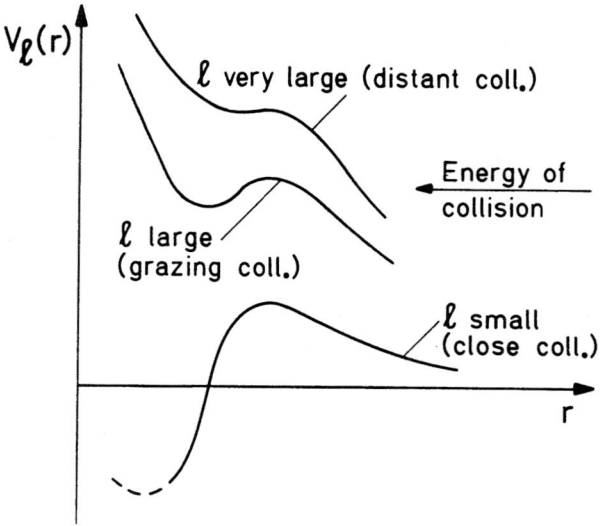

Fig. 7.1 Close, grazing and distant collisions above the Coulomb barrier

$$(C_1 + n) + C_2 \rightarrow C_1 + (C_2 + n)$$
$$A + a \rightarrow b + B \quad.$$

$$(7.1.1)$$

It is our aim to use classical arguments, especially conservation of linear and angular momentum, to obtain constraints on the angular momenta and Q-values for which the reaction takes place with maximum intensity. We choose the z-axis perpendicularly to the scattering plane and denote by $Y_{\ell_1 \lambda_1}(\theta_1, \varphi_1)$ and $Y_{\ell_2 \lambda_2}(\vartheta_2, \varphi_2)$ the orbital angular momentum wave functions of the nucleon n in the nuclei A and B, respectively, with $\vec{r}_1$ ($\vec{r}_2$) the c.m. coordinate of the nucleon in nucleus A(B), see fig. 7.1.1 . We neglect the spin of the transferred nucleon, and assume that its mass M is much smaller than the reduced mass m of the particles A and a (only first-order terms in M/m are considered in the following treatment). Moreover, we consider reactions well above the Coulomb barrier so that the relative velocity of the two nuclei in the interaction region does not change significantly in the transfer process. We denote by R_1 and R_2 the radii of the nuclei b and a , respectively. Simple arguments can be used to derive the conditions for a maximal value of the transfer probability. These qualitative arguments can be made more quantitative by

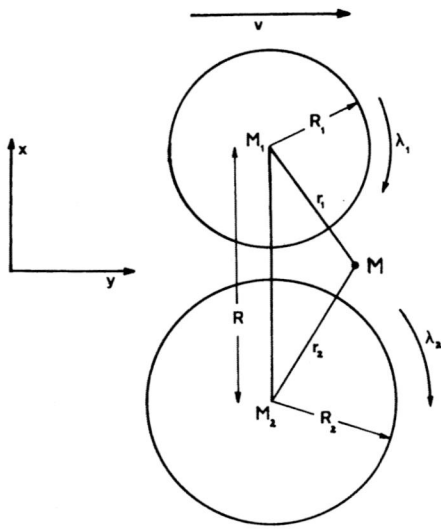

Fig. 7.1.1 Diagram of the transfer process. (From [BR 72a])

using the semiclassical theory, but this is not done here. The conditions are

(i) Conservation of the y-component of the momentum of the transferred particle.
This yields

$$Mv - \frac{\hbar\lambda_1}{R_1} = \frac{\hbar\lambda_2}{R_2} \quad , \tag{7.1.2}$$

if we assume that the transfer is a surface reaction which takes place at the point
of closest contact between a and b . Actually, because of the uncertainty relation
we cannot require the momentum transfer $\hbar\Delta k$ to be exactly zero. We estimate the
uncertainty in the y-direction to be D , that in the x-direction to be ΔR and find
by approximating $\Delta R \approx R_1 \approx R_2 \approx \frac{1}{2}R = D$,

$$\hbar\Delta k = \left| Mv - \frac{\hbar\lambda_1}{R_1} - \frac{\hbar\lambda_2}{R_2} \right| \leq max\left\{ \frac{2\hbar}{R} , 2\frac{\hbar}{R}|\lambda_1 - \lambda_2| \right\} . \tag{7.1.3}$$

(ii) Conservation of the z-component of the total angular momentum. The z-component
of the angular momentum of relative motion is given by $mv(R_1 + R_2) = mvR$. We thus
have

$$\hbar(\lambda_2 - \lambda_1) + \delta(mvR) = 0 . \tag{7.1.4}$$

The change $\delta(mvR)$ has contributions from changes in m, in v, and in R. The change in m is given by

$$\delta m = - \frac{M_1(M_2+M) - M_2(M_1+M)}{M_1+M_2+M} = \frac{M(M_2-M_1)}{M_1+M_2+M} \; . \tag{7.1.5}$$

The change in R is given by

$$\delta R = \left(R_1 + R_2 + \frac{M}{M_1}R_1\right) - \left(R_1 + R_2 + \frac{M}{M_2}R_2\right) = M\left(\frac{R_1}{M_1} - \frac{R_2}{M_2}\right) \tag{7.1.6}$$

where we have only taken into account the change in the relative coordinate before and after the collision. Combining these results, we obtain

$$\delta(mvR) \approx \frac{1}{2}Mv\,(R_1 - R_2) + \frac{R}{v}\,\delta\left(\frac{1}{2}mv^2\right) \; . \tag{7.1.7}$$

The change $\delta\left(\frac{1}{2}mv^2\right)$ in the kinetic energy of relative motion is just the Q-value of the reaction. We write $\delta\left(\frac{1}{2}mv^2\right) = Q_{eff}$ and put $Q_{eff} = Q$ for neutron transfer, and $Q_{eff} = Q - (e^2/R)(Z_1 - Z_2)$ for a proton, where the additional term comes from the change in the Coulomb interaction between the nuclei. Combining all these results, we find as the second condition

$$\Delta \ell_z = \lambda_2 - \lambda_1 + \frac{1}{2}k\,(R_1 - R_2) + \frac{RQ_{eff}}{\hbar v} \approx 0 \; . \tag{7.1.8}$$

(iii) The transfer probability is largest when the transferred nucleon is near the reaction plane, i.e., when $\vartheta_1 \approx \vartheta_2 \approx \frac{\pi}{2}$ in $Y_{\ell_1\lambda_1}, Y_{\ell_2\lambda_2}$. Since $Y_{\ell_i\lambda_i}\left(\frac{\pi}{2},\varphi\right) = 0$ unless $\lambda_i + \ell_i$ is even, we find the conditions

$$\lambda_1 + \ell_1 \quad \text{even} \qquad \text{and} \qquad \lambda_2 + \ell_2 \quad \text{even} \; . \tag{7.1.9}$$

The conditions (i) to (iii) lead to a selectivity of the reaction. Knowing ℓ_1 and ℓ_2 as well as Q_{eff}, one tries to choose values for λ_1 and λ_2 in accord with condition (iii) such that conditions (i) and (ii) are met. Generally speaking, this is possible for some states populated in the transfer reaction and not for others. The resulting selectivity of the transfer process is in good agreement with experimental results. Some values of λ_1 , λ_2 and $\Delta\ell_z$, Δk extracted from the data on reactions $^{12}C(^{12}C,^{11}C)^{13}C$ (neutron transfer) and $^{12}C(^{12}C,^{9}Be)^{15}O$ (^{3}He transfer) are shown in table 7.1.1 . Those transitions are strongest which have the smallest values for $|\Delta\ell_z|$ and $R|\Delta k|/2$. According to [SC 72] , the ratio of cross-sections populating the $5/2^+$ and $7/2^+$ states in ^{15}O, respectively, is roughly 1/4, in nice qualitative agreement with table 7.1.1 .

Table 7.1.1

Characteristic values for favoured one-nucleon transfers in $^{12}C(^{12}C,^{11}C)^{13}C$ and $^{12}C(^{12}C,^{9}Be)^{15}O$ (from [BR 72a])

States	Q(MeV)	λ_1	ℓ_2	λ_2	$\|\Delta\ell_z\|$	$\frac{1}{2} R \|\Delta k\|$
^{13}C						
$1/2^-$, g.s.	-13.7	-1	1	1	1.2	2.2
$5/2^+$, 3.8 MeV	-17.5	-1	2	2	1.0	1.2
^{15}O						
$5/2^+$, 5.2 MeV	-19.4	1	2	2	3.5	3.6
$7/2^+$, 7.3 MeV	-21.4	1	4	4	1.9	1.6

7.2   Semiclassical description [BR 72b, BR 72c, MA 73, BR 74a, KN 75]

An improvement over the classical description is obtained if one uses the semiclassical approximation. Just as in the theory of Coulomb excitation, this means that the trajectories of the fragments are introduced, and the motion of the fragments along these trajectories is treated classically, while the transfer of nucleons is described quantum-mechanically. In the semiclassical approach, two independent steps are taken. The first one concerns the calculation of the reaction amplitudes for a given classical trajectory. It involves the fundamental semiclassical approximation and seems to be rather insensitive to quantum-mechanical corrections. The second step uses the results of the first in the evaluation of cross-sections. It is sensitive to quantum-mechanical effects, but can be carried out in what has

been called a semiquantal treatment in which no reference is made to classical concepts. This point was mentioned already when the semiclassical limit of the quantal
theory of Coulomb excitation was discussed in chapter 6 . Because of the semiclassical
limit this description is valid only for the case of weak absorption (cf. section 5.1).

We consider the reaction A + a → B + b . The relative motion of the two
centers of mass is treated classically and is obtained by solving the equation of
motion containing the combined Coulomb and nuclear potentials. Let s be a channel
index specifying the states of nuclei a and A , and let the corresponding wave
functions be $\psi_a(\xi_a)$ and $\psi_A(\xi_A)$ where the ξ's are a set of internal coordinates.
Neglecting antisymmetrization, we introduce the wave functions

$$\psi_s(\xi) = \psi_a(\xi_a)\, \psi_A(\xi_A) \tag{7.2.1}$$

and expand the total wave function Ψ_E in the form

$$\Psi_E = \sum_s d_s(\vec{r})\, \psi_s(\xi) \tag{7.2.2}$$

where $\vec{r}$ is the distance between the centers-of-mass of a and A . Note that $\vec{r}$
is strictly independent of s. only if s refers to the same fragmentation, but
different intrinsic excitations of a , or A , respectively. For the application to
transfer reactions, it is thus important that the mass transfer is small in comparison
to both the masses of a and A . The function Ψ_E is the stationary solution of

$$H\Psi_E = E\Psi_E \quad , \tag{7.2.3}$$

with H given by

$$H = T(\vec{r}) + V_{int}(\xi,\vec{r}) + H_o(\xi) \quad . \tag{7.2.4}$$

Here, $H_o(\xi)$ is the sum of the intrinsic Hamiltonians of a and A , so that

$$H_o(\xi)\, \psi_s(\xi) = \varepsilon_s\, \psi_s(\xi) \quad , \tag{7.2.5}$$

$T(\vec{r})$ is the kinetic energy operator of relative motion, and $V_{int}(\vec{\varsigma}, \vec{r})$ is the inter-action potential between internal and relative coordinates.

In deriving the semiclassical approximation, we follow the work of Knoll and Schaeffer $\begin{bmatrix} KN \ 75 \end{bmatrix}^{\dagger}$. For simplicity of presentation, we confine ourselves to the consideration of one-dimensional motion and replace the vector $\vec{r}$ by the coordinate x. We introduce the classical action function $S(x)$ as the solution of the equation

$$\left\{ \frac{1}{2m} \left(S'(x) \right)^2 + U(x) - \bar{E} \right\} = 0 \ . \tag{7.2.6}$$

Here, $S'(x) = \frac{d}{dx} S(x)$ and $U(x)$ and $\bar{E}$ are auxiliary quantities. It is our purpose in the following to describe the phase change of $d_s(x)$ by $S(x)$ and in this way to obtain the semiclassical approximation. Subsequently, $U(x)$ and $\bar{E}$ are chosen in such a way that the conditions for the semiclassical approximation are met as well as possible. We accordingly write

$$d_s(x) = c_s(x) \ exp \left(\frac{i}{\hbar} S(x) \right) \ . \tag{7.2.7}$$

After a simple calculation, we find, using eq. (7.2.6),

$$T(x) d_s(x) = exp\left[\frac{i}{\hbar} S(x) \right] \left\{ -\frac{i\hbar}{m} S'(x) c_s'(x) - \frac{\hbar^2}{2m} c_s''(x) + \left[-\frac{i\hbar}{2m} S''(x) + \bar{E} - U(x) \right] c_s(x) \right\}. \tag{7.2.8}$$

It is consistent with the classical equation (7.2.6) defining $S(x)$ if we neglect in the last curly bracket of eq. (7.2.8) the term $-\frac{i\hbar}{2m} S''(x)$ in comparison with $\bar{E} - U$. A straightforward calculation shows that this is possible, if

$$\bar{\chi}(x) \ll \frac{|\bar{E} - U(x)|}{|U'(x)|} \ . \tag{7.2.9}$$

Here, $\bar{\chi}(x) = \frac{\hbar}{p(x)}$ with $p = S'(x)$, and the condition (7.2.9) is seen to be the usual condition of the classical approximation, i.e. that the wave length be small in comparison with the characteristic length over which the potential changes. Assuming

---

$\dagger$ cf. also subsection 2.4.1

this condition to be met, we insert the expansion (7.2.2) into eq. (7.2.3) and use

eqs. (7.2.4) to (7.2.6). Multiplying the resulting equation from the left with $\psi_t(\xi)$

and integrating over ξ , we obtain

$$i\hbar \left(\frac{1}{m} S'(x) \right) c_s'(x) + \frac{\hbar^2}{2m} c_s''(x) = (\bar{E} - E + \varepsilon_s - U) c_s(x)$$
$$+ \sum_t \langle \psi_s | V_{int} | \psi_t \rangle c_t(x) . \tag{7.2.10}$$

We have assumed that V_{int} is local in x , so that it commutes with S(x). We now

observe that $\frac{1}{m} S'(x) = v(x) = \frac{dx}{dt}$. We introduce the classical trajectory x=x(t)

and accordingly rewrite eq. (7.2.10) in the form

$$i\hbar \dot{c}_s(t) = \sum_r \left\{ (\bar{E} - E + \varepsilon_s - U(t)) \delta_{sr} + \langle \psi_s | V_{int} | \psi_r \rangle \right\} c_r(t)$$
$$= \sum_r A_{sr}(x(t)) c_r(t) . \tag{7.2.11}$$

The last equation defines the matrix $A_{sr}(x(t))$. In passing from eq. (7.2.10) to

eq. (7.2.11), we have neglected $\frac{\hbar^2}{2m} c_s''$ in comparison with $i\hbar v c_s'$. This is

justified if $\frac{\hbar^2}{2m} |c_s''| \ll \hbar v |c_s'|$. With the help of eq. (7.2.11), this condition

takes the form

$$\frac{\hbar}{2m\dot{x}^2} \left| \sum_r \left(\dot{A}_{sr} + \frac{1}{i\hbar} (A^2)_{sr} \right) c_r(t) \right| \ll \left| \sum_r A_{sr} c_r(t) \right| \tag{7.2.12}$$

which is fulfilled if

$$\hbar \left| \sum_r A'_{sr} c_r \right| \ll 2p \left| \sum_r A_{sr} c_r \right| \tag{7.2.13a}$$

and

$$\left| \sum_r (A^2)_{sr} c_r \right| \ll 2m\dot{x}^2 \left| \sum_r A_{sr} c_r \right| \tag{7.2.13b}$$

both hold. Eq. (7.2.13a) can be cast in a form like eq. (7.2.9) and thus implies

that λ must be very small in comparison with the length over which the matrix A

changes significantly. Hence, V_{int} must have this property in addition to U , for

which the corresponding condition is formulated in eq. (7.2.9). Since λ is in-

versely proportional to m , this condition can always be met for heavy ions of
sufficiently high energy. For instance, for Ar + Hg at 270 MeV, we find with a
Coulomb barrier of approximately 160 MeV, from eq. (2.1) $\lambda_{cB} \approx 0.15 \, fm$. This has to be
compared with a typical length of $\approx$ 1 fm (nuclear diffuseness!) over which the
potential changes significantly. To fulfill condition (7.2.13b), we must choose
U(x(t)) appropriately. Obviously, a good choice would be one which makes the norm of
A_{ik} small in comparison with $2m\dot{x}^2$. This is impossible, however, since A_{ik} is not
bounded (the ε_s extend all the way up to infinity). We therefore require instead
that A_{ik} , weighted with those c_i which are significantly different from zero,
should be small in comparison with $m\dot{x}^2$:

$$\left| \sum_{s,r} c_s^*(t) \, A_{sr} \, c_r(t) \right| \ll 2m\dot{x}^2 . \tag{7.2.14}$$

The left-hand side can be worked out very easily. Using $\sum_r |c_r|^2 = 1$ (this follows
from eq. (7.2.11) and the initial condition $\lim\limits_{t \to -\infty} c_r(t) \exp(i\varepsilon_r t/\hbar) = \delta_{ro}$)
we find

$$\left| \bar{E} - E - \mathcal{U}(x(t)) + \sum_s \varepsilon_s |c_s|^2 + \sum_{s,r} c_s^*(t) \langle \psi_s | V_{int} | \psi_r \rangle \, c_r(t) \right| \ll 2m\dot{x}^2 . \tag{7.2.15}$$

For physical reasons, we want to choose $\bar{E} = E$ so that for large negative times, the
trajectory described by S(t) and the one actually followed by the projectile coincide.
This suggests putting

$$\mathcal{U}(x(t)) = \sum_s \varepsilon_s |c_s|^2 + \sum_{r,s} c_s^*(t) \langle \psi_s | V_{int} | \psi_r \rangle \, c_r(t) . \tag{7.2.16}$$

With this choice of U , the left-hand side of the inequality (7.2.15) actually
vanishes. The choice (7.2.16), suggested by the semiclassical approximation, is
physically very reasonable. The potential U(x) is the sum of the total intrinsic
excitation energy, and of the interaction energy. The latter vanishes whenever the
two ions after the collision attain a sufficiently large distance. The same statement
does not apply to $\sum_s \varepsilon_s |c_s|^2$ and, hence, to U(x(t)). We see that U(x(t)) automatically

takes account of the loss of energy of relative motion due to intrinsic excitation.

In summary, we have obtained the following coupled equations.

$$m \ddot{x} = - \frac{d}{dx} U(x) \; , \tag{7.2.17a}$$

$$i\hbar \dot{c}_s(t) = (\varepsilon_s - U(x)) c_s(t) + \sum_r < \psi_s | V_{int} | \psi_r > c_r(t) \; , \tag{7.2.17b}$$

$$U(x) = \sum_s \varepsilon_s | c_s(t) |^2 + \sum_{s,r} c_s^*(t) < \psi_s | V_{int} | \psi_r > c_r(t) \; . \tag{7.2.17c}$$

The last of these defines U , the second is equivalent to eq. (7.2.11), and the first is the classical equation of motion which is equivalent to eq. (7.2.6). For the construction of $\frac{d}{dx} U(x)$ it is important to notice that

$$\frac{d}{dx} U(x) = \sum_{s,r} c_s^*(t) < \psi_s | \frac{\partial}{\partial x} V_{int} | \psi_r > c_r(t) \; . \tag{7.2.17d}$$

This relation follows from the definition (7.2.17c) and from the eqs. (7.2.17b). Eqs. (7.2.17) can be simplified further by introducing the amplitudes

$$\tilde{c}_s(t) = \exp \left\{ \frac{i}{\hbar} \int_{-\infty}^{t} dt' \, U(x(t')) \right\} c_s(t) \; . \tag{7.2.18}$$

These amplitudes obey the coupled equations

$$m \ddot{x} = - \sum_{s,r} \tilde{c}_s^*(t) < \psi_s | \frac{\partial}{\partial x} V_{int} | \psi_r > \tilde{c}_r(t) \; , \tag{7.2.19a}$$

$$i\hbar \dot{\tilde{c}}_s(t) = \varepsilon_s \tilde{c}_s(t) + \sum_r < \psi_s | V_{int} | \psi_r > \tilde{c}_r(t) \; . \tag{7.2.19b}$$

The system of eqs. (7.2.19) can be solved without explicit construction of the potential U(x). A similar, but somewhat more involved, derivation can be given for the coupled equations describing the three-dimensional motion of $\vec{r}(t)$ [KN 75] .

We now turn to the second step, i.e. the evaluation of cross-sections in terms of the amplitudes $\tilde{c}_s(+\infty)$. The simplest approximation consists in writing

$$\left(\frac{d\sigma}{d\Omega}\right)_{s \to r} = |\tilde{c}_r(+\infty)|^2 \sqrt{\left(\frac{d\sigma}{d\Omega}\right)_s^{cl} \left(\frac{d\sigma}{d\Omega}\right)_r^{cl}} \qquad (7.2.20)$$

where $\left(d\sigma/d\Omega\right)_r^{cl}$ is the classical cross-section for elastic scattering, given by

eq. (2.1.12), and where $\tilde{c}_r(+\infty)$ is the solution of eq. (7.2.19b) subject to the

initial condition $\lim\limits_{t \to -\infty} \tilde{c}_r(t) \exp(i\varepsilon_r t/\hbar) = \delta_{rs}$. The time-dependence of

$\langle \psi_s | V_{int} | \psi_r \rangle$ has to be calculated by taking for $x = x(t)$ a suitable

classical trajectory. This procedure is known to give an accurate result for in-

elastic scattering well below the Coulomb barrier, where only one trajectory con-

tributes to each deflection angle. At and above the Coulomb barrier, there are always

quantum-mechanical effects. These are caused by (i) the interference of two or

more trajectories, (ii) by diffraction (cf. sections 5.1, 5.3 and 5.4), and

(iii) by absorption. Effects caused by (i) and (ii) can be taken into account in

terms of a suitable generalization of the equations (7.2.19) which we do not describe

[KN 75] . By solving the coupled system of equations, one obtains simultaneously

the trajectory $\vec{r}(t)$ and the amplitudes $\tilde{c}_s(t)$. The latter can be used in eq.(7.2.20).

A much more accurate expression for the cross-section is obtained, however,

if one interprets $\hat{c}_r = \lim\limits_{t \to \infty} \{c_r(t) e^{i\varepsilon_r t/\hbar}\}$ as reaction amplitudes for the transition

$s \to r$. This can be done by introducing $\ell_s = bk_s$ where b is the impact para-

meter and $\hbar k_s$ the initial momentum in the entrance channel, and by writing

$\hat{c}_{\ell_s \to r}$ instead of $\hat{c}_r$. The scattering amplitude then takes the form

$$f_{s \to r}(\theta,\phi) = \frac{1}{i\sqrt{k_s k_r}} \sum_{\ell_s=0}^{\infty} \sqrt{2\ell+1} \exp[i(\delta_s+\delta_r)] \, \hat{c}_{\ell_s \to r} \, Y_{\ell_s \mu}(\theta,\phi) . \qquad (7.2.21)$$

Here, μ is the total change in the magnetic quantum number in the reaction

$s \to r$, and δ_s and δ_r are the elastic phase shifts in channels s and r ,

respectively. The expression (7.2.21) contains a major part of diffraction effects

because it is not restricted by the stationary phase approximation (cf. section 5.4).

It is easy to verify that for large values of ℓ_s , eq. (7.2.21) leads back to the

classical limit (7.2.20). Eq. (7.2.21) has the form of an elastic scattering amplitude. This shows that it holds only if the angular momentum $\Delta\ell$ transferred in the
reaction is small compared to the ℓ-values which give the main contributions to the
sum (7.2.21). This is always the case in heavy-ion grazing collisions ($\Delta\ell \ll \ell_{gr}$).

Various effects not discussed above can also be examined critically. These
are the overlap of channel wave functions, antisymmetrization effects, and in particular recoil effects which are quite important in HI reactions for energies above
the Coulomb barrier. The incorporation of these in the semiclassical approximation
has been considered in [BR 72b]. The conditions derived in section 7.1 with
additional correction terms have been obtained in the semiclassical description
[BR 72a].

Absorption effects can be incorporated into the semiclassical approximation.
One writes [BR 72b, BR 74 b], cf. eq. (5.1.29),

$$\tilde{c}_s'(t) = \tilde{c}_s(t) \, exp \left\{ \frac{1}{\hbar} \int_{-\infty}^{t} W_s \left(r(t') \right) dt' \right\} \tag{7.2.22}$$

where $\tilde{c}_s(t)$ obey the coupled semiclassical equations without absorption, and where
the exponential contains the integral over the absorptive part of the potential
along the classical trajectory, this leading to an exponential damping of the wave
packet. The form (7.2.22) can be shown to be adequate for weak absorption. For
strong absorption the semiclassical treatment breaks down because of diffraction.

The agreement obtained in some calculations is exemplified in fig. 7.2.1 ,
where a comparison between the semiclassical calculation and a full-fledged first-
order quantum-mechanical calculation is shown. Absorption causes the cross-section
to decrease monotonically with increasing Θ for $\Theta \geq 75°$. Quantum-mechanical calculations as well as complex trajectory calculations [MA 75] lead within $\approx 10\%$ to
the same result. While the agreement is impressive, the above mentioned difficulty
expected for strong absorption suggests caution in the use of the semiclassical
approximation for inelastic scattering, unless the improved versions formulated
recently [KN 75, MA 75] are used.

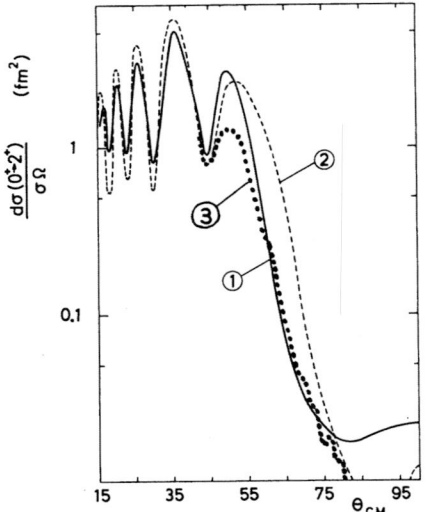

Fig. 7.2.1 Inelastic scattering of ^{16}O (60 MeV) on ^{58}Ni, leading
to the first excited (2^+) state of ^{58}Ni. The solid curve
(1) is the semiclassical, the dashed curve (2) the quantum-
mechanical result. The parameters of the calculation are
the same as in fig. 5.1.9 . (From [MA 73]). The dotted
curve (3) is obtained by taking absorption into account
according to eq. (7.2.22) [BR 74a] .

7.3   Coupled channels and DWBA

In the preceding section 7.2 , we have discussed the inelastic and transfer
processes in the semiclassical approximation, i.e., by treating the relative motion
of the colliding nuclei classically. The applicability of the semiclassical approximation is restricted to high energies and/or very heavy colliding nuclei. In many
cases which have been studied experimentally, these conditions are not satisfied,
so that a quantum-mechanical description is necessary. We shall discuss here the
coupled-channels formalism for inelastic and transfer processes.

7.3.1   Coupled channels for inelastic scattering [TA 65]

We introduce the coupled-channels formalism for the simplest case, i.e., the
inelastic scattering of an inert projectile by a target nucleus. The Hamiltonian is
then given by

$$H(\vec{r}, \xi) = T(\vec{r}) + h(\xi) + V(\vec{r}, \xi) \qquad (7.3.1)$$

where T is the kinetic energy of relative motion, $h(\xi)$ is the Hamiltonian of the target nucleus and $V(\vec{r}, \xi)$ describes the interaction between the projectile and the target. Suppose the target wave-functions, determined by

$$h(\xi) \, \Phi_{nIM_I}(\xi) = \varepsilon_{nI} \, \Phi_{nIM_I}(\xi) \tag{7.3.2}$$

with spin I, magnetic quantum number M_I and the remaining quantum numbers n are known. Introducing the eigenfunctions of total angular momentum by

$$\mathcal{Y}_{n(\ell I) JM}(\hat{r}, \xi) \equiv \sum_{m_\ell, M_I} (\ell m_\ell I M_I | JM) \, Y_{\ell m_\ell}(\hat{r}) \, \Phi_{nIM_I}(\xi) \tag{7.3.3}$$

we expand the scattering solution for given total angular momentum J, M and parity π in the form

$$\Psi_{JM,\pi}(\vec{r}, \xi) = \frac{1}{r} \sum_\alpha g_\alpha^{J\pi}(r) \, \mathcal{Y}_{\alpha JM,\pi}(\hat{r}, \xi) \tag{7.3.4}$$

where $\alpha \equiv n(\ell, I)$ is referred to as (reduced) channel index. Note that due to rotational invariance of the total Hamiltonian, $g_\alpha^{J\pi}(r)$ does not depend on M. Inserting this expansion for Ψ into the Schrödinger equation and making use of the orthonormality of the functions $\mathcal{Y}_{\alpha JM,\pi}(\hat{r}, \xi)$, we obtain for each total angular momentum J and parity π a set of coupled equations

$$\left(T_\alpha + V_{\alpha\alpha}^{J\pi}(r) - E_\alpha \right) g_\alpha^{J\pi}(r) = - \sum_{\beta \neq \alpha} V_{\alpha\beta}^{J\pi}(r) g_\beta^{J\pi}(r) \tag{7.3.5}$$

for the radial functions $g_\alpha^{J\pi}(r)$. Here, $T_\alpha \equiv \hbar^2/(2m) [-d^2/dr^2 + \ell(\ell+1)/r^2]$ and $E_\alpha \equiv E - \varepsilon_\alpha$. For a pure Coulomb interaction we obtain eqs. (6.4.2) and (6.4.3) for the description of the Coulomb excitation.

If the c.m. energy is larger than the Coulomb barrier, the two particles can penetrate each other. In the description of these processes through the eqs. (7.3.5) one encounters the following problems.

(i) As compared to Coulomb excitation below the Coulomb barrier much more states (including compound states) become available in these processes.

(ii) The projectile cannot be considered any more as an inert particle. One has to worry about the non-orthogonality of the wave functions in the region of the overlap. The exchange of nucleons cannot be neglected in general. If projectile and target differ only by a few nucleons, the exchange of particles may become especially important (cf. subsection 7.3.6).

Effects from (i) and (ii) are usually taken into account approximately by introducing a (phenomenological) optical potential.

7.3.2 Optical potential in the coupled-channels formalism

We discuss here the introduction of the optical potential as given by Glendenning [GL 69] . In section 4.1 , the optical potential for elastic scattering has been defined. We generalize now the definition of the optical potential to the case of several strongly coupled channels. Following Feshbach [FE 62] , we introduce the two projection operators P and Q with P + Q = 1 . The operator P projects onto the channels we want to take explicitly into account in the coupled-channels formalism. Projecting the Schrödinger equation

$$(E - H)(P + Q)\Psi = 0 \qquad (7.3.6)$$

onto Q we find

$$(E - QHQ)(Q\Psi) = QHP(P\Psi) \qquad (7.3.7)$$

which can be solved formally by

$$Q\Psi = (E - QHQ + i\varepsilon)^{-1} QHP(P\Psi) . \qquad (7.3.8)$$

Here the infinitesimal quantity $i\varepsilon$ with $\varepsilon > 0$ has been introduced because only outgoing waves are allowed in $Q\Psi$. Projection of eq. (7.3.6) on P and substitution of eq. (7.3.8) leads to

$$\{ PHP + PHQ(E - QHQ + i\varepsilon)^{-1} QHP - E \}(P\Psi) = 0 \qquad (7.3.9)$$

where the first two terms can be interpreted as an effective Hamiltonian in the P space. Introducing H in the simple form of eq. (7.3.1), we get for the effective Hamiltonian

$$H_{eff}(\vec{r}, \xi; E) = T(\vec{r}) + Ph(\xi)P + V_{GOMP}(\vec{r}, \xi; E)$$ (7.3.10)

with the generalized optical-model potential (GOMP)

$$V_{GOMP}(\vec{r}, \xi; E) \equiv PV(\vec{r}, \xi)P + PVQ\,(E - QHQ + i\varepsilon)^{-1}QVP.$$ (7.3.11)

Eq. (7.3.9) still describes the exact solution in the subspace P . A simplification arises because of the finite resolution of experimental detection. Therefore, only the cross-section

$$\left\langle \frac{d\sigma_{\vec{\beta}\vec{\alpha}}}{d\Omega_{\beta}} \right\rangle_I \propto \langle |T_{\vec{\beta}\vec{\alpha}}|^2 \rangle_I = |\langle T_{\vec{\beta}\vec{\alpha}} \rangle_I|^2 + \langle |T_{\vec{\beta}\vec{\alpha}} - \langle T_{\vec{\beta}\vec{\alpha}} \rangle_I|^2 \rangle_I$$ (7.3.12)

averaged over an energy interval I is measured. The index $\vec{\alpha}$ denotes the channel quantum numbers n, I, M_I (cf. eq.(7.3.2)) and the relative wave vector $\vec{k}_\alpha$. The fluctuation part (second term on the r.h.s.) can be evaluated approximately within the statistical model (cf. section 8.1). The direct part (first term on the r.h.s.) defines the optical potential (cf. section 4.1) in the following way: Assume a Lorentz distribution for calculating the mean value of $T_{\vec{\beta}\vec{\alpha}}$, then we have (cf. [ME 62, eq.(XIX.20)])

$$\langle T_{\vec{\beta}\vec{\alpha}}(E) \rangle_I = T_{\vec{\beta}\vec{\alpha}}(E + iI) = \langle \Phi_{\vec{\beta}}(E+iI)| V_{GOMP}(E+iI)|\Psi_{\vec{\alpha}}^{(+)}(E+iI)\rangle.$$ (7.3.13)

The main contribution to the integral (7.3.13) comes from the interaction region. In this region the plane wave $\Phi_{\vec{\beta}}(E+iI) \approx \Phi_{\vec{\beta}}(E)$ for $I \ll E$. For the same reason the scattering solution $\Psi_{\vec{\alpha}}^{(+)}(E+iI)$ determined by eq. (7.3.9) for $E \rightarrow E + iI$ is approximated by $\Psi_{\vec{\alpha}}^{(+)opt}(E)$ which is the solution of the same eq. (7.3.9) with E replaced by E + iI in V_{GOMP} only. Thus the optical-model wave function $\Psi_{\vec{\alpha}}^{(+)opt}(E)$ is the solution of the Schrödinger equation in the subspace P with the optical potential

$$\mathcal{V}(E) \equiv V_{GOMP}(E + iI) .$$ (7.3.14)

It can be shown that the error introduced by these approximations in eq. (7.3.13),
is negligibly small if E exceeds the Coulomb barrier by more than ≈ 2 MeV. The
averaged T-matrix approximated in this way,

$$< T_{\vec{\beta}\vec{\alpha}} (E) >_I \approx < \phi_{\vec{\beta}} (E) \mid \mathcal{V}(E) \mid \Psi_{\vec{\alpha}}^{(+)opt} (E) > \qquad (7.3.15)$$

can be read also directly from the asymptotic behaviour of $\Psi_{\vec{\alpha}}^{(+)opt}(E)$. An expansion into eigenfunctions of total angular momentum leads to the coupled channel
eqs. (7.3.5) with V replaced by the optical potential $\mathcal{V}$(E) . Thus the averaged
T-matrix is determined by the solution of the coupled channel eqs.(7.3.5) in the
subspace P .All effects on the averaged T-matrix from the compound states and other
channels which are contained in Q are accounted for by the replacement of the bare
interaction V by the optical potential $\mathcal{V}$(E) . In comparison to subsection 4.1
the optical model for the elastic channel has been generalized to the treatment of
several strongly coupled channels (subspace P).

The optical-model potential is, according to eqs. (7.3.11) and (7.3.14),
given by the matrix elements

$$\mathcal{V}_{\alpha\beta} (E) = V_{\alpha\beta} + \sum_r \frac{V_{\alpha r} V_{\beta r}^*}{E - E_r + iI} + \sum_q \int dE_q \frac{V_{\alpha q} V_{\beta q}^*}{E - E_q + iI} \qquad (7.3.16)$$

for $\alpha, \beta \in$ P   and zero for α and/or $\beta \in Q$. For $P \equiv |\alpha\rangle\langle\alpha|$ the optical potential
reduces to the elastic optical potential treated in section 4.1 . If there are many
resonances in the energy region between $E - iI$ and $E + iI$, and if the matrix
elements $V_{\alpha r}$ are random numbers , then the sum over r gives a considerable contribution only for α = β . The same argument holds for the sum over q if the
number of open channels is large. Then, the non-diagonal elements of $\mathcal{V}$(E) contain
a much smaller contribution from the last two terms than the diagonal elements. Thus,
the coupling between the channels is given almost completely by the "bare" inter-
action $V_{\alpha\beta}$ (real coupling) if $I \gg |V_{\alpha r}|, |V_{\beta r}|, |V_{\alpha q}|$ and $|V_{\beta q}|$. Significant
contributions from the sum over r and q to the coupling between two channels are
expected (i) if there is a strong channel-channel correlation via the compound states

(e.g. in the case of isobaric resonances) or (ii) if there is a strong direct coupling to a state or channel which is not taken explicitly into account. In the latter case it is advantageous to treat this channel explicitly in the coupled-channels formalism. Before applying the coupled-channels formalism to the scattering of α-particles by deformed nuclei, we shall first consider the distorted-wave Born approximation.

7.3.3 Distorted-wave Born approximation (DWBA)

In the distorted wave Born approximation the diagonal elements of $\mathcal{V}(E)$ are treated exactly whereas the non-diagonal elements are treated in first-order perturbation theory. The diagonal element of $\mathcal{V}$ lead to the distortion of the plane waves in the channels. These distorted waves are determined by

$$(T_\alpha + \mathcal{V}_{\alpha\alpha} - E_\alpha)\, \chi_{\vec{\alpha}}^{(\pm)} = 0 . \tag{7.3.17}$$

According to ref. $\left[\text{ME 62, eq. (XIX. 74)}\right]$ the T-matrix element (7.3.14) in the DWBA is given by

$$T_{\vec{\beta}\vec{\alpha}}^{DWBA} = \langle \chi_{\vec{\beta}}^{(-)} | \mathcal{V} | \chi_{\vec{\alpha}}^{(+)} \rangle \tag{7.3.18}$$

for $\beta \neq \alpha$ where the scattering solution $\Psi_{\vec{\alpha}}^{(+)opt}$ has been replaced by the corresponding distorted wave $\chi_{\vec{\alpha}}^{(+)}$. This amounts to a first-order Born approximation with respect to the non-diagonal elements of $\mathcal{V}$. Thus the excitation of the target is considered only as a one-step process, which is usually illustrated by the diagram of fig. 7.3.1 . We point out that the use of a complex coupling potential $\mathcal{V}_{\vec{\alpha}\vec{\beta}}$ formally takes into account also effects from multiple-step processes which lead over intermediate states of the Q space. This is obvious from the discussion following eq. (7.3.16).Computer codes for DWBA calculations have been developed (e.g. DWUCK) and frequently applied to HI reactions $\left[\text{AU 70}\right]$. In cases where the coupling of different channels is too strong, the coupled-channels equations have to be used.

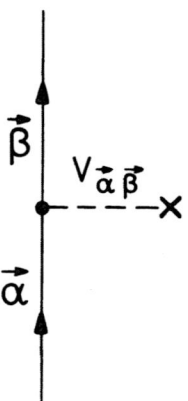

Fig. 7.3.1   Illustration of a one-step process as described by
the DWBA.

7.3.4   Alpha particle scattering from deformed nuclei

As one of the most impressive applications of the coupled-channels formalism, we discuss here the inelastic scattering of α-particles from deformed nuclei.
The rotational motion of the deformed nucleus is strongly coupled to the α-particle
motion by the phenomenological nuclear interaction

$$\mathcal{V}_N(r-R) = -\frac{V_0 + i\,W_0}{1 + exp\,\dfrac{r-R}{a}} \tag{7.3.19}$$

where

$$R = r_0\,A^{1/3}\left[1 + \sum_\lambda \beta_\lambda\,Y_{\lambda 0}(\theta',\phi')\right] \tag{7.3.20}$$

describes the surface of the deformed nucleus with respect to the body-fixed
coordinates θ', ϕ' . The imaginary part is assumed to account for other intrinsic
excitations (besides rotations) of the deformed nucleus and of compound-nucleus
formation. The interaction (7.3.19) is expected to be the same for all channels of

the ground-state rotational band. In order to simplify the evaluation of the coupling matrix elements, the nuclear interaction is expanded (cf. [GL 69]) into a sum of multipole operators. The coupling matrix elements are calculated by introducing the Bohr-Mottelson rotational wave functions [BO 53] . We do not want to go into details of the calculations but rather state the approximations which lead to the theoretical results discussed below:

(i)   The target nucleus is assumed to be well deformed in order that the rotational states are well described by the Bohr-Mottelson wave functions. A microscopic theory has been given by Sevgen [SE 72] which leads essentially to the same results for strongly deformed nuclei.

(ii)   The incident energy is assumed to be much higher than the excitation energies of the rotational states.

(iii)   Band mixing is considered to be small, so that the restriction to the ground state band is justified.

In fig. 7.3.2 the main features of the theoretical analyses of the inelastic scattering of alpha particles from deformed nuclei are illustrated. The example which is studied is the excitation of the 2^+, 4^+, 6^+ levels of the ground state rotational band of ^{154}Sm by 50 MeV alpha particles. In part (a) the best fit obtained within the coupled-channels formalism is shown. The curves reproduce the experimental data very well. Part (b) illustrates the influence of the hexadecapole deformation. Note that the 4^+ and 6^+ excitations are extremely sensitive to the value of β_4 . Part (c) demonstrates the failure of the DWBA. A much larger absorption is needed to get the best fit to the 2^+ cross-section. In addition very different deformation parameters are extracted. The inelastic cross-section for exciting the 4^+ and 6^+ states cannot be fitted at all. Part (d) illustrates the effect of Coulomb excitations on the cross-section.

Similar experiments at smaller incident energies (14 to 18 MeV alpha particles) on ^{152}Sm [BR 73] also lead to conclusive results on deformation parameters. Also lighter deformed nuclei like ^{20}Ne and ^{28}Si [RE 73] have been studied. The results for the deformation parameters are $\beta_2 = 0.42$, $\beta_4 = 0.29$ for ^{20}Ne and $\beta_2 = -0.39$,

- 198 -

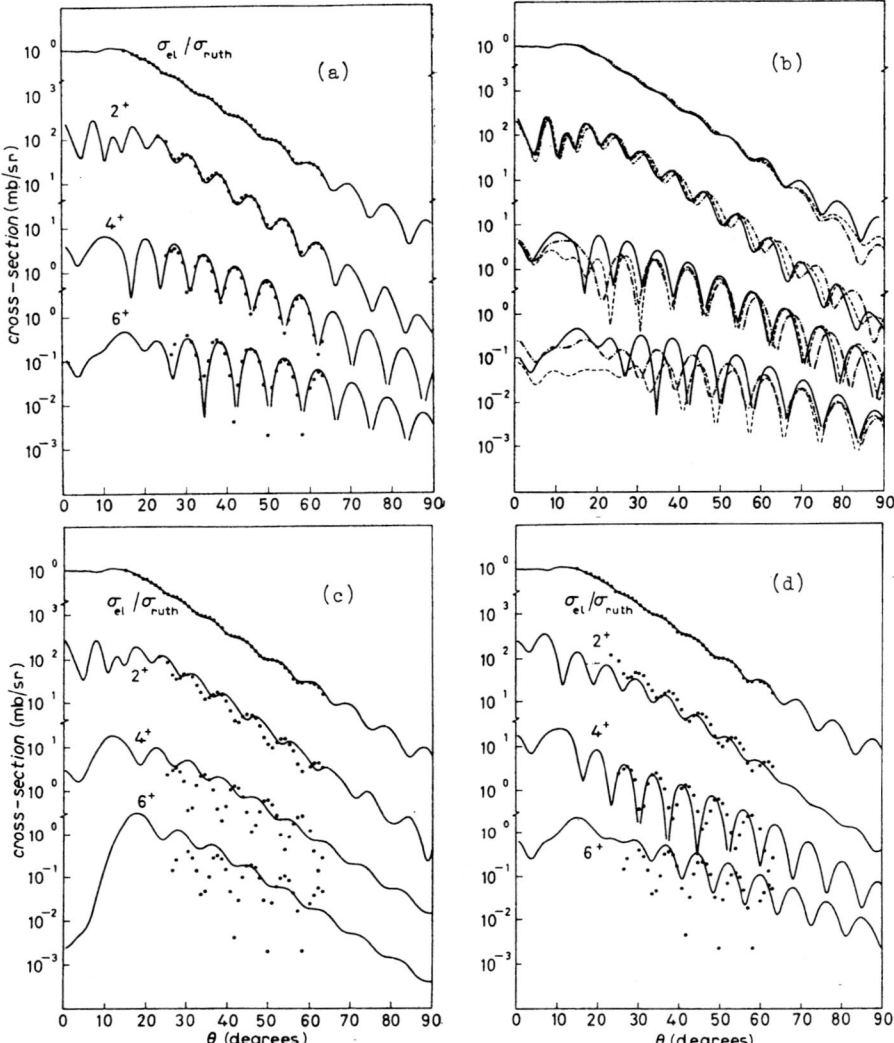

Fig. 7.3.2 Differential cross-sections for the excitation of the lowest
ground-state rotational levels of ^{154}Sm by 50 MeV alpha
particles (from [GL 69]). (a) Best fit CC calculations
with V_o = 65.9 MeV, W_o= 27.3 MeV, r_o = 1.440 fm, a = 0.637 fm,
β_2 = 0.225, β_4 = 0.05. (b) Sensitivity of the cross-sections
to the value of β_4 (———·——· β_2 = 0.235, β_4 = -0.05; - - -
β_2 = 0.235, β_4 = 0; ——— β_2 = 0.225, β_4 = 0.05). (c) Best fit
DWBA calculations with V_o= 34.6 MeV, W_o= 29.4 MeV, r_o= 1.404 fm,
a = 0.819 fm, β_2 = 0.3, β_4 = 0.15, β_6 = 0.075 . (d) Coulomb
excitation neglected.

$\beta_4 = 0.27$ for ^{28}Si .

7.3.5   Generalization of the coupled-channels formalism to transfer reactions

We generalize the coupled-channels formalism of subsection 7.3.1 to include also transfer reactions by using in the expansion (cf. eq. (7.3.4))

$$\Psi_{JM\pi} = \sum_{\alpha} \frac{1}{r_{\alpha}} g_{\alpha}^{J\pi}(r_{\alpha}) \, y_{\alpha JM\pi}(\hat{r}_{\alpha}, \xi_{\alpha}) \qquad (7.3.21)$$

also channels which correspond to different reaction products. Now $\alpha = aA, \, n(\ell,I)$ includes the numbering of the transfer channels aA, bB, cC, ... There are two important additional complications in comparison with the coupled-channels formalism for inelastic scattering. These are the non-orthogonality and the overcompleteness of the basis states $y_{\alpha JM\pi}$. Practically, the overcompleteness is no real difficulty, because the coupled-channels calculation can only be done for a rather limited number of channels. The non-orthogonality introduces a more severe problem [OH 70, GO 72] . In principle it is possible to define a biorthogonal set of basis states by

$$\int \tilde{y}_{\alpha}^{*}(\hat{r}_{\alpha}, \xi_{\alpha}) \, y_{\beta}(\hat{r}_{\beta}, \xi_{\beta}) \, d\hat{r}_{\alpha} d\xi_{\alpha} = \delta_{\alpha\beta} \qquad (7.3.22)$$

in order to get the same form for the coupled equations as before, i.e.,

$$(T_{\alpha} + V_{\tilde{\alpha}\alpha} - E_{\alpha}) \, g_{\alpha}(r_{\alpha}) = -\sum_{\beta \neq \alpha} \int \tilde{y}_{\alpha}^{*} \, V \, y_{\beta} \, g_{\beta} \, \frac{r_{\alpha}}{r_{\beta}} \, d\hat{r}_{\alpha} d\xi_{\alpha} . \qquad (7.3.23)$$

It has been shown [GO 72] that effects due to non-orthogonality are negligibly small in some typical cases. Therefore, it is usually assumed that one can neglect the non-orthogonality and simply regard the basis states of eq. (7.3.21) to be orthogonal.

The coupled-channels equations (7.3.23) contain the coupling to all orders between the channels which are explicitly introduced. Since the numerical solution of the coupled equations is very time-consuming, one has carefully to choose the channels and look for approximations. We shall discuss briefly the coupled-channels Born approximation and the coupled reaction channels formalism.

(i) The coupled-channels Born approximation (CCBA) has been introduced by Ascuitto and Glendenning [AS 69, 72] in connection with transfer reactions on deformed nuclei. To be specific, we consider the A(p,t)B reaction on even-even deformed nuclei where the strong coupling between the rotational states has to be taken into account. The CCBA consists in solving the coupled eqs. (7.3.5) separately for the (pA) and (tB) rotational channels with the appropriate asymptotic behaviour and then calculating the T-matrix elements in the Born approximation with respect to the coupling between (pA) and the (tB) channels. This means that the two-particle transfer is treated in Born approximation whereas the coupling between the channels of the ground-state rotational band is treated to all orders for tB as well as for pA. This is illustrated in fig. 7.3.3 where the first-order coupling for the transfers is indicated by the thin arrows and the all-order coupling for the inelastic transitions within the ground-state rotational bands by the double arrows. Thus the CCBA can be considered as a generalization of the DWBA to the inclusion of the rotational degrees of freedom in the distorted waves. In the numerical treatment of the CCBA, Ascuitto and Glendenning simplify the calculation considerably by using the solution for the (pA) channels for generating source terms via the transfer

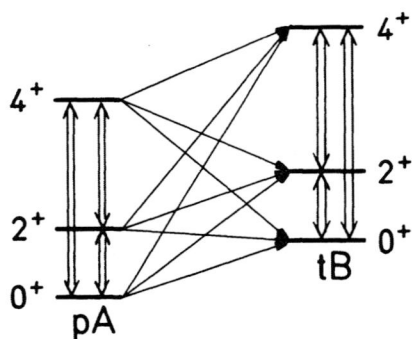

Fig. 7.3.3 Illustration of the coupled-channel Born approximation
(CCBA).

matrix elements in the coupled equations for the (tB) channels. By this procedure
they avoid calculating the general solutions for the (tB) channels and the additional
calculation of the DWBA matrix element. Fig. 7.3.4 illustrates the relative importance
of various transfers leading to the 2^+ rotational state in the reaction ^{176}Yb(pt)^{174}Yb.
Note that the direct and indirect transfers are comparable in magnitude. Fig. 7.3.5
gives the final result of the CCBA calculation as compared with DWBA calculation.
The partly destructive interference of the different contributions of fig. 7.3.4
leads to the final result which is a significant improvement over the DWBA calculation
given in fig. 7.3.5 . Note that even the elastic cross-section is affected considerably by the strong coupling between the rotational states.

(ii)   The coupled reaction channels formalism (CRC) has been introduced by Coker
et al. [CO 73] in order to study reactions which proceed through an intermediate
fragmentation cC different from both the initial aA and final bB fragmentation (cf.
fig. 7.3.6). The authors can show that in the case of ^{48}Ca(h,t)^{48}Sc(0^+IAS) with an
intermediate α-channel, it is sufficient to iterate the coupled equations (7.3.23)

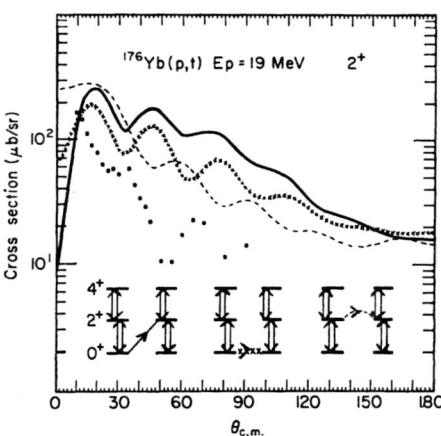

Fig. 7.3.4   Illustration of various transfer contributions leading to
the 2^+ rotational state in ^{174}Yb(p,t). The three curves
correspond to the transitions indicated in the lower part
of the figure. (From [AS 72]).

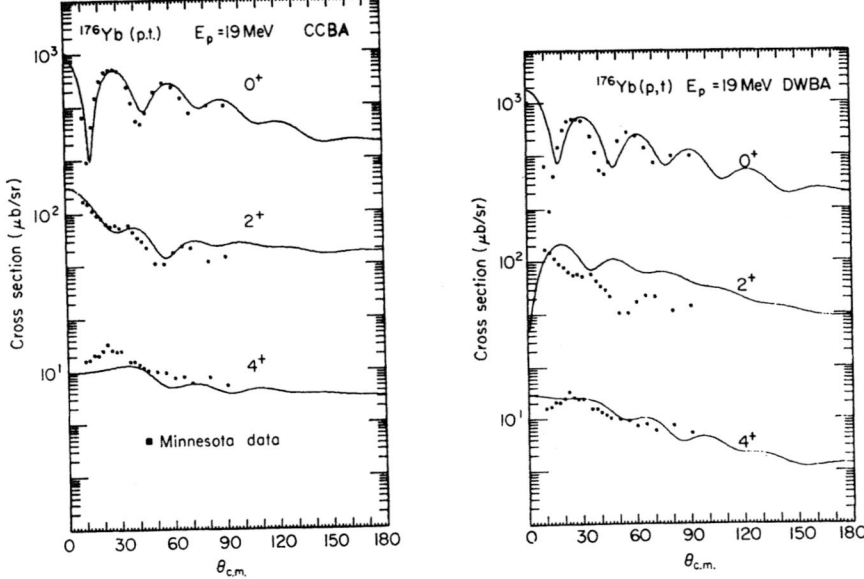

Fig. 7.3.5 CCBA and DWBA fits to the ^{176}Yb(p,t) reactions leading to
the 0^+, 2^+ and 4^+ levels of the ground-state rotational
band of ^{174}Yb. (From [AS 72]).

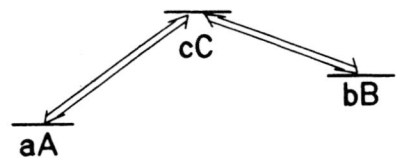

Fig. 7.3.6 Illustration of the coupled reaction channels formalism .

up to second order (2nd order Born approximation). We do not want to go into details of these calculations. The results are at present not very conclusive, mainly because of the uncertainties in the coupling matrix elements. Effects connected with the non-orthogonality of the basis states, the recoil correction (cf. subsection 7.3.7) and the zero-range approximation in calculating the r.h.s. of eq. (7.3.23) might also be important.

7.3.6   Elastic transfer

As a simple application of the coupled-channels formalism, we consider in this subsection the elastic scattering between two nuclei A and B $\equiv$ (A + b) which differ only by one or a few nucleons (denoted by b). The elastic scattering amplitude can be considered to consist of two components, corresponding to the "direct" elastic scattering

$$(A_1 + b) + A_2 \longrightarrow (A_1 + b) + A_2 \tag{7.3.24}$$

and to the elastic transfer reaction

$$(A_1 + b) + A_2 \longrightarrow A_1 + (A_2 + b) \tag{7.3.25}$$

which cannot be distinguished from the direct elastic scattering, because the two cores A_1 and A_2 are identical. In atomic physics, such a process (like H^++ H) is usually called a "resonant particle transfer", cf. [MO 65, GE 69, MC 70] . We prefer to call this process "elastic transfer", because the term "resonant" is somewhat misleading (actually, no resonances in the atomic and nuclear cross-sections are observed). The elastic transfer is a good example for a reaction in which the exchange properties of indistinguishable particles become relevant. In addition, the elastic transfer seems to be one of the best understood heavy-ion transfer reactions where multistep processes can be studied. The following discussion is concerned (i) with a qualitative discussion of the elastic transfer, (ii) with the basic theory of the elastic transfer of a particle bound in states with $j \le \frac{1}{2}$, (iii) with generalizations and applications. For a recent review see [OE 75] .

(i) Qualitative considerations

We consider the scattering of A on B $\equiv$ (A + b) and assume for simplicity that

the particle b (e.g. an α-particle) is bound to the spin-zero core in a state of

(total) angular momentum j = 0 . In a semiclassical picture, cf. fig. 7.3.7 , there

are two trajectories which contribute to the elastic cross-section, the direct

scattering (7.3.24) and the transfer process (7.3.25). Therefore, the scattering

amplitude consists of the sum of a direct and a transfer part

$$f_{el}(\theta) = f_d(\theta) + f_{tr}(\pi - \theta)$$ (7.3.26)

and the cross-section consists of three terms,

$$\left(\frac{d\sigma}{d\Omega}\right)_{el} = \left|f_d(\theta)\right|^2 + \left|f_{tr}(\pi - \theta)\right|^2 + 2\left|f_d(\theta)\right|\left|f_{tr}(\pi - \theta)\right|\cos\left[\phi_d(\theta) - \phi_{tr}(\pi - \theta)\right]$$ (7.3.27)

where $\phi_d(\theta)$ and $\phi_{tr}(\pi - \theta)$ denote the phases of the corresponding amplitudes.

The direct scattering contribution $\left|f_d(\theta)\right|^2$ usually shows the typical diffraction

behaviour (cf. sections 5.3. and 5.4) which is illustrated in fig. 7.3.8 . The pure

transfer part is peaked around the grazing angle and usually shows rather weak os-

cillations which are determined by the grazing angular momentum. The interference

between the direct and the transfer contribution gives rise to strong oscillations

of the elastic cross-section. In order to obtain a rough estimate for the period of

these oscillations, we assume that the angular dependence of the phases of $f_d(\theta)$

and $f_{tr}(\pi - \theta)$ are given by the corresponding Coulomb phases, cf. $\left[\text{ME 61, ch. XI §7}\right]$.

With $\theta = \frac{\pi}{2} + \delta$ we then get

$$\phi_d(\theta) - \phi_{tr}(\pi - \theta) \approx const + 2\eta\delta$$ (7.3.28)

for $\delta \ll 1$, i.e., for $\theta \approx \frac{\pi}{2}$. Hence, the period of the oscillations which are

characteristic for the elastic transfer is

$$\delta_o \approx \frac{\pi}{\eta}$$ (7.3.29)

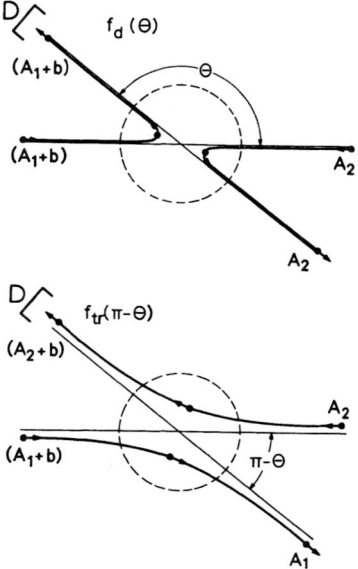

Fig. 7.3.7 Semiclassical picture of (a) the direct and (b)
the transfer process in the elastic scattering
of $(A_1 + b)$ on A_2.(From $\begin{bmatrix} BO\ 74 \end{bmatrix}$).

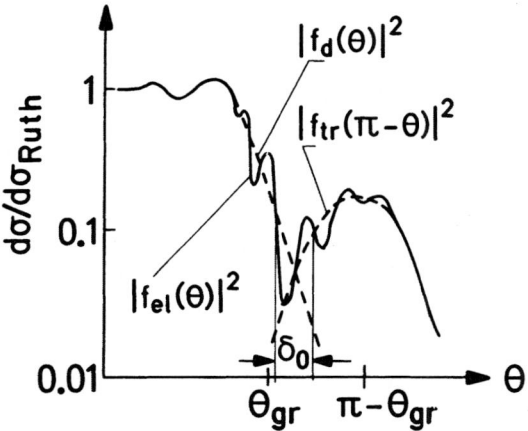

Fig. 7.3.8 Schematic diagram of the elastic cross-section which
illustrates the direct and transfer contributions and
their mutual interference.

around $\Theta = \pi/2$. The same period for the interference oscillations is observed in the scattering of two identical charged nuclei (Mott scattering). The phenomenon of elastic transfer has been first observed and interpreted as an interference of direct and transfer scattering amplitudes by Gobbi et al. $\begin{bmatrix} GO\ 68 \end{bmatrix}$ for $^{13}C + {}^{12}C$.

(ii) Basic theory of the elastic transfer (LCNO model)

We give here a non-perturbative treatment of the elastic transfer $\begin{bmatrix} OE\ 70,73 \end{bmatrix}$ although a DWBA can very well be applied to several cases (cf. part (iii) of this subsection). We consider the transfer of a valence particle b which is allowed to carry a spin, between two identical cores A_1 and A_2 which have zero spins. The Hamiltonian is assumed to be

$$H = T(\vec{r_1}) + V_{opt}(r_1) + h(\vec{r_1}, \xi_2) = T(\vec{r_2}) + V_{opt}(r_2) + h(\vec{r_2}, \xi_1) \qquad (7.3.30)$$

where $\vec{r_i}$ is the vector connecting the center of the core A_i (with i = 1 or 2) with the center of the nucleus B_j (with i ≠ j = 2 or 1) as illustrated in fig. 7.3.9 .

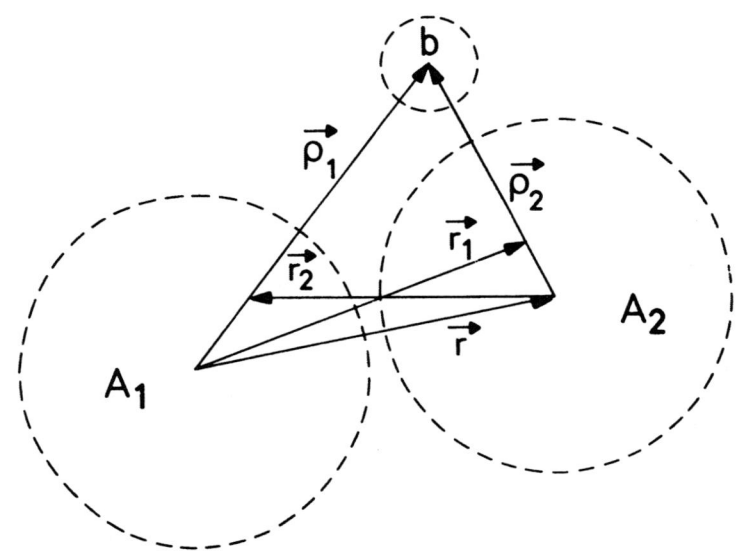

Fig. 7.3.9 Coordinates used to describe the relative positions of A_1, A_2 and b .

The quantity $\vec{\xi}_i \equiv \{\vec{\rho}_i, \vec{s}\}$ denotes the spin coordinate $\vec{s}$ and the space coordinate $\vec{\rho}_i$ of the valence particle taken with respect to the core A_i ($i = 1$ or 2). The operators $T(\vec{r}_i) = \frac{\hbar^2}{2m} \Delta_{\vec{r}_i}$ and $V_{opt}(r_i)$ are the kinetic energy and the optical potential of the relative motion of A_i and $B_{j \neq i}$. The imaginary part in V_{opt} takes account of inelastic and compound elastic processes which are not treated explicitly. The Hamiltonian of the valence particle is given by

$$ h(\vec{r}_j, \xi_i) = t(\vec{\rho}_j) + v(\xi_i) + v(\xi_j) \qquad (7.3.31) $$

where $t(\vec{\rho}_j)$ and $v(\xi_j)$ denote the kinetic and potential energy of the relative motion of b and A_j .

In solving the Schrödinger equation with the Hamiltonian given by eqs. (7.3.30) and (7.3.31) we consider the expansion

$$ \Psi = \frac{1}{\sqrt{2}} \sum_\alpha \left\{ \tilde{F}_\alpha^{(1)}(\vec{r}_1) \phi_\alpha(\xi_1) + \tilde{F}_\alpha^{(2)}(\vec{r}_2) \phi_\alpha(\xi_2) \right\} \qquad (7.3.32) $$

where the wave functions $\phi_\alpha(\xi_i)$ describe the bound states of the valence particle b in the potential $v(\xi_i)$ of the core A_i . The wave functions $\tilde{F}_\alpha^{(i)}(\vec{r}_i)$ describe the relative motion of A_i and $B_{j \neq i}$. The expansion (7.3.32) represents the solution of the Schrödinger equation in the subspace spanned by the ϕ_α's and would lead to a set of coupled equations similar to eq. (7.3.23), cf. [IM 75] .

We consider here for simplicity only the elastic transfer of a valence particle which is bound in a single-particle state ϕ_α with total angular momentum $j \leq \frac{1}{2}$. Then the coupled eqs. (7.3.23) can be decoupled under the following two approximations: (a) For small enough energies and/or for valence particle mass small compared to the core mass, the wave length of $\tilde{F}_\alpha^{(i)}(\vec{r}_i)$ becomes large as compared to the difference $|\vec{r}_i - \vec{r}|$, $\vec{r}$ being the vector connecting the centers of the cores. Hence, $\vec{r}_1$ and $\vec{r}_2$ can be replaced in the wave functions $\tilde{F}_\alpha^{(i)}(\vec{r}_i)$ of relative motion by $\vec{r}$ and $-\vec{r}$, respectively. This approximation amounts to neglecting the recoil effects (cf. subsection 7.3.7) and is justified for c.m.

energies small compared to $30 \cdot A^{1/3}$ MeV [OE 73] where A is the mass number of the cores. (b) We restrict the expansion (7.3.32) to the ground states, only. Thus we assume that the coupling to the excited states of the valence particle is not too large and can be accounted for by the optical potential. Of course, the coupling to excited states is not small in the interior region. But since the absorption is very strong there, the wave functions $\tilde{F}_\alpha^{(i)}(\vec{r}_i)$ of relative motion are already strongly damped, so that the strong coupling to the excited states does not influence the asymptotic behaviour of $\tilde{F}_\alpha^{(i)}(\vec{r}_i)$.

Because of the identity of the cores, it is convenient to introduce the linear combination of nuclear orbitals (LCNO)

$$\Phi_m^{(p)} = \frac{1}{\sqrt{2}} \left[\phi_m(\xi_1) + p\,\phi_m(\xi_2) \right] \qquad (7.3.33)$$

with the symmetry $p = \pm 1$ with respect to the exchange of the cores. For simplicity, only the component m of the total angular momentum j with respect to the (laboratory) z-axis is taken to label the wave functions. Then, the expansion (7.3.32) can be written in the form

$$\Psi = \sum_{m,p} F_m^{(p)}(\vec{r})\, \Phi_m^p(\xi_1, \xi_2) \quad . \qquad (7.3.34)$$

Since we are dealing with spin-zero cores (Bosons) the total wave function Ψ has to be invariant with respect to the exchange of the cores. Hence, $F_m^{(p)}(-\vec{r}) = p\,F_m^{(p)}(\vec{r})$ contains only even $(p = +1)$ or odd $(p = -1)$ partial waves. Inserting the expansion (7.3.32) into the Schrödinger equation, we find by introducing $\Phi_m^{(p)}$ and $F_m^{(p)}$ according to eqs. (7.3.33) and (7.3.34) for $\vec{r}_1 \to \vec{r}$ and $\vec{r}_2 \to -\vec{r}$

$$\left[T(\vec{r}) + V_{opt}(r) + K(r) + p\,\mathcal{J}(r) - E \right] F_m^{(p)}(\vec{r}) = 0 \quad , \qquad (7.3.35)$$

i.e., uncoupled equations for different p and m . Here, we have neglected the overlap between $\phi_m(\xi_1)$ and $\phi_m(\xi_2)$ which gives a higher order correction to the quantities

$$K(r) \equiv \int d\xi \, \phi_m^*(\xi_1) \, v(\xi_2) \, \phi_m(\xi_1) \ , \quad \mathcal{J}(r) \equiv \int d\xi \, \phi_m^*(\xi_1) \, v(\xi_2) \, \phi_m(\xi_2) \ , \quad (7.3.36)$$

i.e., the direct and exchange parts of the interaction due to the presence of the valence particle. Both $K(r)$ and $\mathcal{J}(r)$ are independent of m (for $j \leq 1/2$). The important feature of eq. (7.3.35) is aside from its being a decoupled equation, the symmetry dependence of the exchange potential term $p \, \mathcal{J}(r)$. Since $F_m^{(+)}(\vec{r})$ consists only of even partial waves and $F_m^{(-)}(\vec{r})$ only of odd partial waves, even and odd partial waves are scattered from different potentials, the difference being $2 \mathcal{J}(r)$. This is illustrated in fig. 7.3.10 for $\mathcal{J}(r) > 0$. For energies close to the Coulomb barrier, the even partial waves are more strongly absorbed than the odd partial waves. The result is an even-odd staggering in the values of $|S_\ell|$ as illustrated in fig. 7.3.11 . The asymptotic behaviour of the functions $F_m^{(p)}(\vec{r})$ determines the scattering amplitudes $f_p(\theta)$. The differential cross-section then turns out to be

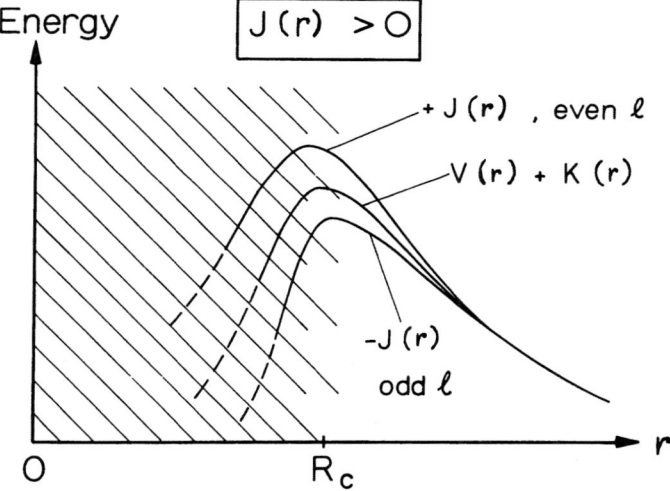

Fig. 7.3.10 Schematic diagram of the real potential energies with splitting due to the exchange potential for $\mathcal{J}(r) > 0$. The shaded area indicates the region of strong absorption due to the imaginary part of the optical potential. (From [OE 73]).

$$\frac{d\sigma}{d\Omega} = \left| f_+(\theta) + f_-(\theta) \right|^2 \tag{7.3.37}$$

where $f_+(\theta)$ and $f_-(\theta)$ contain only even or odd partial waves, respectively. The interference of even and odd partial waves leads to the oscillations in the differential cross-section shown in fig. 7.3.12 for $^{12}C + ^{13}C$. The case $\mathcal{I}(r) < 0$ has been included there in order to demonstrate the dependence of the interference pattern on the sign of $\mathcal{I}(r)$.

It is useful to make connection to our qualitative discussion leading to the separation of the elastic scattering amplitude into a direct and a transfer contribution, eq. (7.3.26). The corresponding separation is obtained in the quantum-mechanical description given here, by using the symmetry property $f_p(\theta) = p \, f_p(\pi - \theta)$. Then the scattering amplitude can be written in the form of eq. (7.3.26) with

$$\begin{aligned} f_d(\theta) &= \tfrac{1}{2} \left[f_+(\theta) + f_-(\theta) \right] , \\ f_{tr}(\theta) &= \tfrac{1}{2} \left[f_+(\theta) - f_-(\theta) \right] . \end{aligned} \tag{7.3.38}$$

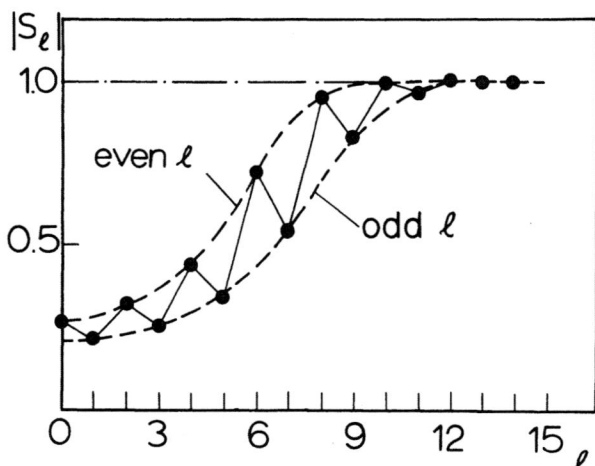

Fig. 7.3.11 Schematic diagram of the reflection coefficient $\eta_\ell \equiv |S_\ell|$ as function of ℓ showing the even-odd staggering for $\mathcal{I}(r) > 0$. (From [OE 73]).

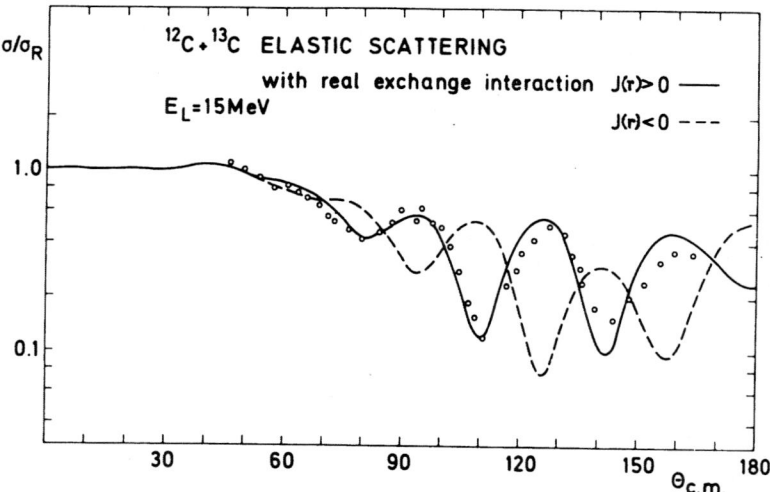

Fig. 7.3.12 Ratio of differential cross-sections to Rutherford
cross-section showing the influence of the sign of
the exchange potential $\mathcal{J}(r)$. (From [OE 75]).

Assuming a polynomial expansion $f_p(\theta)$ into powers of the strength s of the exchange potential $\mathcal{J}(r)$, we have

$$f_{\pm}(\theta) = f_o(\theta) \pm s f_1(\theta) + s^2 f_2(\theta) \pm s^3 f_3(\theta) + \cdots \qquad (7.3.39)$$

where $f_i(\theta)$ denotes the contribution from the i'th order transfer. Hence, the
direct part $f_d(\theta)$ consists only of even powers of s and, therefore, corresponds
to an even number of transfers, whereas the transfer part $f_t(\theta)$ consists only
of odd powers of s and thus corresponds to an odd number of transfers.

(iii) Generalizations and applications of the LCNO model, DWBA calculations

In order to apply the LCNO model, it is necessary to consider some many-body
effects [OE 73] : First of all, the treatment can be generalized in the Hartree-
Fock theory to apply also to the transfer of a hole, of two particles, two holes or
a particle and a hole. In addition, it is shown that correlations are taken into
account to a rather good approximation by multiplying the exchange potential by

the spectroscopic factor S_j of the single-particle state.

With these generalizations, the theory has been applied to various elastic transfer reactions, where the particles or holes are bound in a state $j = 1/2$ or 0 to a spin-zero core. Figs. 7.3.13 and 7.3.14 are typical results. Fig. 7.3.13 shows the elastic transfer of a particle in the scattering of ^{12}C on ^{13}C and the elastic transfer of a hole in the scattering ^{13}C on ^{14}C . The hole transfer is out of phase as compared to the particle transfer because the sign of the exchange potential $J(r)$ for the hole is opposite to the that for the particle. Fig. 7.3.14 shows the elastic transfer of one and two particles for Si isotopes together with the Mott scattering and the reaction ^{30}Si(^{28}Si,^{29}Si)^{29}Si . It is nicely demonstrated that the oscillations have the same period of 9° as given by eq.(7.3.29).

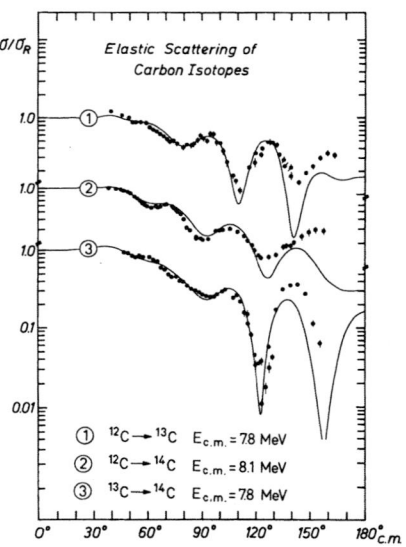

Fig. 7.3.13 Elastic scattering of C isotopes.
(From [OE 73]).

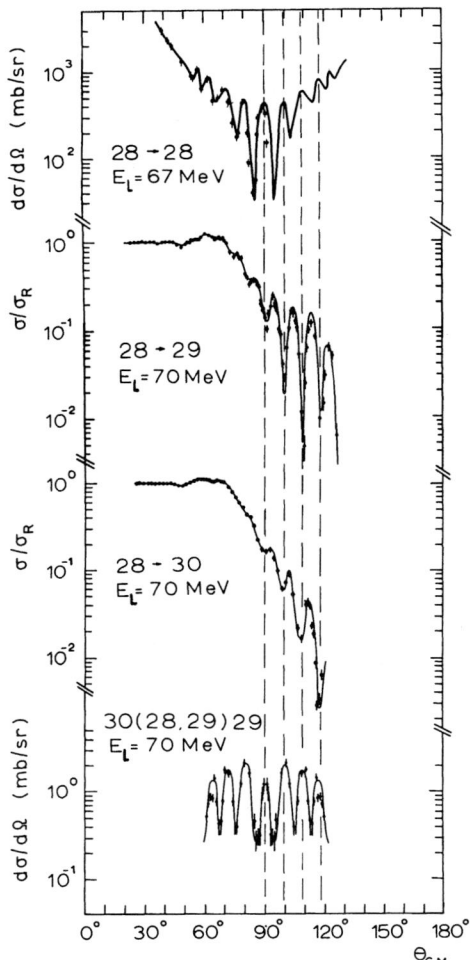

Fig. 7.3.14 Elastic and transfer scattering of Si isotopes.
(From [HI 74]).

Using the LCNO model Baur and Gelbke [BA 73] have shown that, for strongly absorbing potentials, it is possible to treat the exchange potential in first-order perturbation theory, i.e. to calculate the transfer amplitude in the DWBA which yields $f_1(\theta)$ in the expansion (7.3.39) . For $j \leq 1/2$ the DWBA is disadvantageous as compared to the LCNO model because it contains the additional calculation of the DWBA matrix element. But for $j > 1/2$ the LCNO model does no longer lead to decoupled equations, so that the DWBA is of practical value. DWBA calculations have been done for $^{16}O + ^{17}O$ where $j = 5/2$ [GE 73] . It is seen that the transfer of angular momentum becomes important with increasing incident energy. Since the transfer of angular momentum leads to incoherence with respect to the direct elastic scattering, the interference pattern is washed out with increasing energy. An alternative approach [BO 74] has been introduced by using molecular eigenstates for the transferred particle. Then, neglecting Coriolis forces , we again obtain uncoupled equations.   The calculations show that the effect of multiple transfers of the valence nucleon becomes important for shallow optical potentials (V = 17 MeV).

7.3.7   DWBA and recoil effects

In this subsection we are concerned with some technical problems which arise in the calculation of matrix elements occurring on the right-hand side of eq.(7.3.23). The evaluation of these integrals is difficult in the case of transfer reactions because the coordinates are differently defined in each transfer channel. We restrict our discussion to the DWBA matrix element [ME 62, ch. XIX; AU 70]

$$< \chi_f^{(-)} \mid V_f \mid \chi_i^{(+)} > \qquad (7.3.40)$$

for the transfer of a nucleon N between two cores C_1 and C_2 ,

$$(\underbrace{C_1 + N}_{A_1}) + C_2 \longrightarrow C_1 + (\underbrace{C_2 + N}_{A_2}) \,. \qquad (7.3.41)$$

According to the discussion below eq. (7.3.16), the interaction V_f is usually taken as the bare interaction between the particles in the final channel, i.e.,

$$V_f = V_{C_1N}(\vec{\rho}_1) + V_{C_1C_2}(\vec{r})\qquad(7.3.42)$$

where we have used the coordinates of fig. 7.3.15 .

The integral (7.3.40) simplifies significantly if the mass M of the nucleon can be considered to be small as compared to the masses of the cores. Then according to fig. 7.3.15 we have

$$\vec{r}_i \approx \vec{r}_f \approx \vec{r}\qquad(7.3.43)$$

and hence the matrix element of $V_{C_1C_2}$ vanishes if the wave functions of the nucleon N bound to the cores C_1 and C_2, respectively, are considered to be orthogonal (cf. the discussion below eq. (7.3.21)). The matrix element of V_{C_1N} is usually calculated by a method introduced by Buttle and Goldfarb [BU 66, 68] . These authors use an approximate expansion of the nucleon wave function bound to C_2 in terms of wave functions bound to C_1 . Thereby, the transferred angular momentum $\Delta\ell$

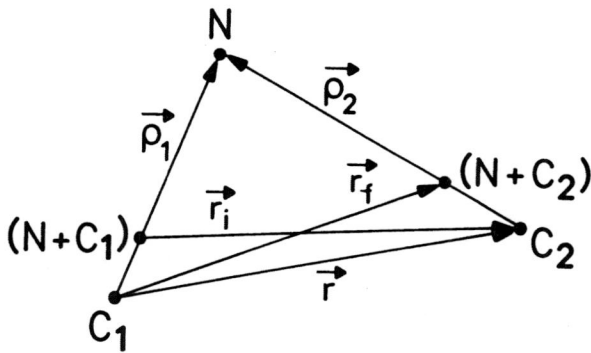

Fig. 7.3.15 Coordinates used for describing the transfer of a nucleon N between two cores C_1 and C_2 .

is introduced which satisfies the triangle relations $|\ell_1 - \ell_2| \leqslant \Delta\ell \leqslant \ell_1 + \ell_2$, $|j_1 - j_2| \leqslant \Delta\ell \leqslant j_1 + j_2$ and the parity conservation $\Delta\ell + \ell_1 + \ell_2$ = even . Here, ℓ_1, ℓ_2, j_1 and j_2 denote the orbital and total angular momenta of the nucleon in the initial and the final state. The transferred angular momenta $\Delta\ell$ characterize the angular distribution of the transfer cross-section.

The approximation (7.3.43) consists in the neglect of recoil effects, because it is strictly valid only in the limit $M \to 0$, i.e., in the limit where the transferred particle carries no momentum. Recoil corrections have been studied by Dodd and Greider [DO 65, GR 66, GR 70] . As qualitatively expected, these recoil effects become significant for energies high above the Coulomb barrier. Two effects arise: (i) With increasing energy the magnitude of the cross-section changes. This effect is due to the mismatch between initial and final angular momentum as discussed in section 7.1 in the classical limit. (ii) With increasing energy the angular distribution is washed out. The reason is that additional values for the transferred angular momenta $\Delta\ell$ not restricted by the parity conservation rule ($\Delta\ell + \ell_1 + \ell_2$ = even) become significant, because the angular momenta have to be taken with respect to the centers of mass and not to the centers of the cores. Methods for approximately calculating recoil corrections in the HI reactions have been developed recently [BR 74b, BA 74] .

7.4   Parametrized phase-shift model in the DWBA for inelastic and transfer
        reactions (Smooth cut-off model and Regge-pole model)

As described in chapter 5 , the parametrization of phase shifts is a useful tool in understanding the elastic HI scattering. The smooth cut-off model, in particular, is capable of treating the diffraction phenomena which are produced by strong absorption (section 5.4). Regge-poles allow for the description of surface waves (quasi-molecular resonances) in an elegant way. It is, therefore, of interest that these models can be successfully applied to direct reactions (inelastic scattering and transfer of nucleons) in HI collisions. For simplicity we restrict our discussion to reactions with no angular momentum transfer ($\Delta\ell = 0$) . Then the inelastic or transfer scattering amplitude is of the form

$$f_{fi}(\theta) = \frac{1}{2ik_f} \sum_{\ell} (2\ell+1) \, f_{\ell}^{(fi)} \, P_{\ell}(\cos\theta) \qquad (7.4.1)$$

where the partial (inelastic or transfer) amplitude is in the recoilless DWBA [BU 66, AU 70]

$$f_{\ell}^{(fi)} \propto \int dr \, F_{\ell}(k_f, r') \, v_0(r) \, F_{\ell}(k_i, r) \quad . \qquad (7.4.2)$$

Here, $F_{\ell}(k_f, r')$ and $F_{\ell}(k_i, r)$ denote the radial distorted waves of the final (wave number k_f) and initial (wave number k_i) channel, respectively. The coupling $v_0(r)$ arises from the interaction between the nuclei by integrating over the internal degrees of freedom. The quantity $r' \equiv r A_{C_2}/A_{C_2+N}$ where A_{C_2} and A_{C_2+N} are the mass numbers of C_2 and C_2+N (cf. eq. 7.3.41). For inelastic scattering r' becomes identical with r .

The essential property of the partial amplitudes $f_{\ell}^{(fi)}$ for HI reactions is the localisation in ℓ space. The lower bound is due to absorption (cf. sections 5.3 and 5.4), the upper bound is due to the finite range of $v_0(r)$. Simple parametrizations of f_{ℓ} have been introduced by Strutinsky [ST 64] and Frahn and Venter [FR 64]. The effect of strong absorption has been discussed by Frahn [FR 73,75] in the smooth cut-off model. An even simpler view has been taken by Fuller and McVoy [FU 74, 75, MC 75] using Regge-poles in order to describe the localisation in ℓ space.

7.4.1   Smooth cut-off model

In order to display explicitly the effects due to absorption, Frahn uses a method of Dar [DA 66] and Sopkovich [SO 62] which is based on the JWKB approximation. The radial wave function $f_{\ell}(k,r)$ is approximated by the regular Coulomb wave function $F_{\ell}(kr)$ multiplied by the square root of the elastic S-matrix elements in the initial and final channels. Hence,

$$f_{\ell} \propto \bar{S}_{\ell}(k_f, k_i) \, I_{\ell}(k_f, k_i) \qquad (7.4.3)$$

where

$$\bar{S}_\ell = \bar{\eta}_\ell \, e^{2i\bar{\phi}_\ell} \equiv \sqrt{S_\ell(k_f') \, S_\ell(k_i)} \qquad (7.4.4)$$

$$I_\ell = \int_0^\infty dr \, \mathcal{F}_\ell(k_f' r) \, v_o(r) \, \mathcal{F}_\ell(k_i r) \qquad (7.4.5)$$

with $k_f' \equiv k_f A_{C_2}/A_{C_2+N}$. In eq. (7.4.3) the effects due to absorption are contained in $\bar{S}_\ell$ and may be parametrized in a way similar to eq. (5.4.2). The sum over ℓ in eq. (7.4.1) is now replaced by an integral over $\lambda = \ell + 1/2$. The main contributions to $f(\theta)$ result from stationary phase points and from λ values where $\bar{\eta}(\lambda)$ is rapidly varying. Similarly to section 5.4 we obtain for weak absorption the semiclassical result (cf. section 7.2) if $I(\lambda)$ is varying smoothly with λ . For strong absorption the diffraction phenomena (Fresnel and Fraunhofer) show up in the angular distribution [FR 73, 75] .

7.4.2   Regge-pole model

The localisation of the partial scattering amplitude $f_\ell^{(fi)}$ in ℓ space is taken into account by introducing a double pole [FU 75] .

$$(2\ell+1) \, f_\ell^{(fi)} = \frac{c}{(\ell-\alpha)^2} \, exp \left\{ 2i \left(\sigma_\ell^{(i)} + \sigma_\ell^{(f)} + \ell \phi \right) \right\} \qquad (7.4.6)$$

where the two poles correspond to the initial and final channel. The four parameters c, Re α, Im α and ϕ are determined by the experimental elastic and inelastic cross-sections. The quantities $\sigma_\ell^{(i)}$ and $\sigma_\ell^{(f)}$ denote the Coulomb phase shifts in the initial and final channel. We note that this parametrization gives a Lorentzian shape for $\eta(\ell)$ with width Im α . Fig. 7.4.1 shows a double-pole fit to the transfer reaction $^{40}Ca(^{13}C,^{12}C)^{41}Ca$ (g.s.) at E_{lab} = 60 MeV and 68 MeV . The success of this parametrization supports the following physical interpretation. A surface wave with angular momentum $\ell \approx$ Re α which is surrounding the nuclear surface dominates the reaction mechanism, the angular distribution becoming insensitive to details of the transfer process.

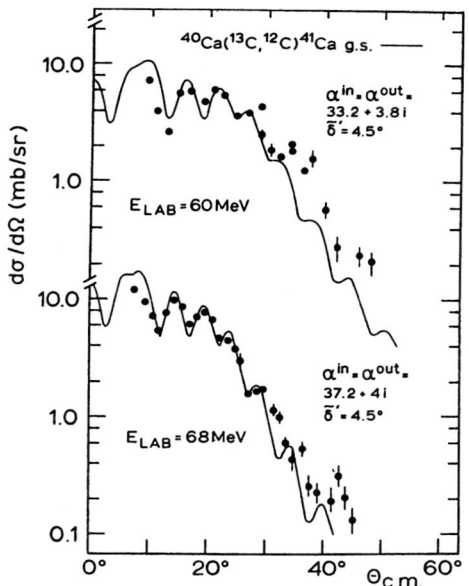

Fig. 7.4.1

Two-Regge-pole fit of the one-particle transfer
reaction ^{13}C(60 MeV, 68 MeV) + $^{40}Ca \rightarrow$ ^{41}Ca(g.s.) + ^{12}C.
Parameters fitted only to the inelastic data.
(From [FU 74]).

References

AS 69   R.J. Ascuitto and N.K. Glendenning, Phys. Rev. 181 (1969) 1396

AS 72   R.J. Ascuitto and N.K. Glendenning, Nucl. Phys. A183 (1972) 60

AU 70   N. Austern, Direct Nuclear Reaction Theories (Wiley Interscience, New York, 1970)

AU 75   N. Austern, Phys. Rev. C12 (1975) 128

BA 73   G. Baur and C.K. Gelbke, Nucl. Phys. A204 (1973) 138

BA 74   A.J. Baltz and S. Kahana, Phys. Rev. C9 (1974) 2243

BO 53   A. Bohr and B.R. Mottelson, Kgl. Dan. Vid. Selsk., Mat.-fys. Medd. 27(1953)No.19

BO 74   H.G. Bohlen and W. Nörenberg, Phys. Lett. 49B (1974) 227

BR 72a D.M. Brink, Phys. Letts. 40B (1972) 37

BR 72b R.A. Broglia and A. Winther, Phys. Rep. 4C (1972) 155

BR 72c R.A. Broglia and A. Winther, Nucl. Phys. A182 (1972) 112

BR 72d R.A. Broglia, S. Landowne and A. Winther, Phys.Lett. 40B (1972) 293

BR 73   W. Brückner, J.G. Merdinger, D. Pelte, U. Smilansky and K. Traxel, Phys.Rev.Lett. 30 (1973) 57

BR 74a R.A. Broglia, S. Landowne, R.A. Malfliet, V. Rostokin and A. Winther, Phys.Rep. 11C (1974) 1

BR 74b P. Braun-Munzinger and H.L. Harney, Nucl.Phys. A223 (1974) 381

BU 66   P.J.A. Buttle and L.J.B. Goldfarb, Nucl.Phys. 78 (1966) 409

BU 68   P.J.A. Buttle and L.J.B. Goldfarb, Nucl. Phys. A115 (1968) 461

CO 73   W.R. Coker, T. Udagawa and H.H. Wolter, Phys. Rev. C7 (1973) 1154

DA 66   A. Dar, Nucl. Phys. 82 (1966) 354

DO 65   L.R. Dodd and K.R. Greider, Phys. Rev. Lett. 14 (1965) 959

FE 62   H. Feshbach, Ann.Phys. (N.Y.) 19 (1962) 287

FR 64   W.E. Frahn and R.H. Venter, Nucl. Phys. 59 (1964) 593

FR 73   W.E. Frahn, in Extended Seminar in Nuclear Physics, Trieste 1973 (IAEA, Vienna, 1975) vol.I, p.157

FR 75   W.E. Frahn, in Classical and Quantum-Mechanical Aspects of Heavy-Ion Collisions, ed. by H.L. Harney et al. (Springer Verlag, Berlin - Heidelberg- New York, 1975) p.102

FU 74   R.C. Fuller and O. Dragun, Phys.Rev.Lett. 32 (1974) 617

FU 75   R.C. Fuller and K.W. McVoy, Phys.Lett. 55B (1975) 121

GE 69   S. Geltman, Topics in Atomic Collision Theory (Academic Press, New York, 1969)

GE 73   C.K. Gelbke, R. Bock, P. Braun-Munzinger, D. Fick, K.D. Hildenbrand, W. Weiss,
         S. Wenneis and G. Baur, Phys. Lett. 43B (1973) 284

GL 69   N.K. Glendenning, Proc. Int. School of Physics "Enrico Fermi", Course XL,
         1967, ed. M. Jean (Academic Press, New York, 1969) p.332

GO 68   A . Gobbi, U. Matter, J.L. Perrenoud and P. Marmier, Nucl.Phys. A112 (1968)537

GO 72   L.J.B. Goldfarb and K. Takeuchi, Nucl.Phys. A181 (1972) 609

GR 66   K.R. Greider and L.R. Dodd, Phys. Rev. 146 (1966) 671

GR 70   K.R. Greider, in Nuclear Reactions Induced by Heavy Ions, ed. R. Bock and
         W.R. Hering (North-Holland Publ. Comp., Amsterdam, 1970) p.217

HA 74   H.L. Harney, P. Braun-Munzinger, and C.K. Gelbke, Z. Physik 269 (1974) 339

HI 74   K.D. Hildenbrand, R. Bock, H.G. Bohlen, P. Braun-Munzinger, D. Fick,
         C.K. Gelbke, U. and W. Weiss, Nucl.Phys. A234 (1974) 167

IM 75   B. Imanishi, H. Ohnishi and O. Tanimura, Phys.Lett. 57B (1975) 309

KN 75   J. Knoll and R. Schaeffer, preprint CEN-Saclay-DPh-T/75/40 (April 1975),
         submitted to Ann.Phys. (N.Y.)

LA 74   S. Landowne and N. Takigawa, Phys. Lett. 50B (1974) 414

MA 73   R.A. Malfliet, Symposium on Heavy-Ion Transfer Reactions, Argonne, 1973
         (Argonne Physics Division, Informal Report PHY-1973) p.605

MA 75   R.A. Malfliet, in Classical and Quantum-Mechanical Aspects of Heavy-Ion
         Collisions, ed. by H.L. Harney et al. (Springer Verlag, Berlin - Heidelberg -
         New York, 1975) p. 86

ME 61   A. Messiah, Quantum Mechanics Vol. I (North-Holland, Amsterdam, 1961)

ME 62   A. Messiah, Quantum Mechanics, Vol. II (North-Holland, Amsterdam, 1962)

MC 70   M.R.C. McDowell and J.P. Coleman, Introduction to the Theory of Ion-Atom
         Collisions (North-Holland, Amsterdam, 1970)

MC 75   K.W. McVoy, in Classical and Quantum-Mechanical Aspects of Heavy-Ion
         Collisions, ed. by H.L. Harney et al. (Springer Verlag, Berlin - Heidelberg -
         New York, 1975) p. 127

MO 65 N.F. Mott and H.S.W. Massey, The Theory of Atomic Collisions (Clarendon Press, Oxford, 1965)

OE 70 W. von Oertzen, Nucl.Phys. A148 (1970) 529

OE 73 W. von Oertzen and W. Nörenberg, Nucl.Phys. A207 (1973) 113

OE 75 W. von Oertzen and H.G. Bohlen, Phys. Reports 19C (1975) 1

OH 70 T. Ohmura, B. Imanishi, M. Ichimura and M. Kawai, Progr.Theoret.Phys. (Kyoto) 44 (1970) 1242

RE 73 H. Rebel and G.W. Schweimer, Z. Physik 262 (1973) 59

SC 72 D.K. Scott, P.N. Hudson, P.S. Fisher, C.U. Cardinal, N. Anyas-Weiss, A.D. Panagiotou, P.J. Ellis, and B. Buck, Phys. Rev. Lett. 28 (1972) 1659

SE 72 A. Sevgen, Nucl.Phys. A195 (1972) 415

SI 71 P.J. Siemens. J.P. Bondorf, D.H.E. Gross and F. Dickmann, Phys.Lett.36B(1971)24

SO 62 N.J. Sopkovich, Nuovo Cimento 26 (1962) 186

ST 64 V.M. Strutinsky, Sov.Phys. JETP 19 (1964) 1401

TA 65 T. Tamura, Revs. Mod. Phys. 37 (1965) 679

8.   Statistical theory

In chapter 4, we have explained why nuclear cross-sections, in the domain where the density of compound states is high, are very rapidly varying functions of energy. This led us to consider _energy-averaged_ cross-sections, with the averaging interval I subject to the conditions (4.2.4), and to the decompositions (4.1.11) and (4.2.5) of the average cross-section into a shape-elastic or direct part, and into a compound-nucleus contribution. These two parts can be viewed as representing the fast and the slow contributions, respectively, to the average cross-section, as explained in the paragraph following eq. (4.1.14). This interpretation leads to the expectation that the shape-elastic or direct part can be described by simple models involving only a few degrees of freedom. The most important of these models were described in chapters 5 and 7 . We now turn to the evaluation of the compound-nucleus contributions to the energy-averaged cross-sections.

A full dynamical description of the complex collision processes which involves the many compound nuclear resonances and, hence, many degrees of freedom of the colliding systems, is very difficult to obtain. However, one is helped by the fact that only the _mean values_ of the cross-sections are of interest. They involve an average over _many_ compound resonances (see condition (4.2.4)) , and one may hope to be able to develop methods which, while being unable to describe the individual behaviour of each resonance or, equivalently, the full cross-section as a function of energy, give a very good account of the average behaviour of cross-sections. It is with this idea in mind that one resorts to a _statistical_ description, analogous to the procedure used in statistical mechanics. Here, the description of the full history of an individual system requires the exact solution of the equations of motion. However, the behaviour of mean values (pressure, temperature, ...) can be predicted reliably without detailed knowledge of such exact solutions. In the spirit of this analogy, one is led to view the collision between two HI and the associated exchange of energy, mass, charge, angular momentum etc. as a dissipative process. If the two ions were sufficiently close to each other for a sufficiently long period of time, this dissipative process might eventually establish complete thermodynamical equilibrium between the two ions, identical with the formation through fusion of a compound nucleus in N. Bohr's picture [BO 36] . In actuality, the time during which the two ions are close to one another is determined by the impact parameter b . For a grazing collision, we expect the processes described in chapters 5 and 7 to dominate the inelastic scattering. Dissipative phenomena are expected to set in as the impact parameter decreases (compare chapter 2.II). However, they can take place only during a short time and will not, in general, lead to complete equilibrium. It is only when the impact parameter approaches zero that fusion may become a likely process. The theoretical description and the experimental analysis of such partial equilibration processes are still in their infancy. However, interesting developments in this field, ranging from the extension of concepts used in the statistical theory of nucleon-induced reactions to the investigation of short-time dissipative phenomena, shock waves, etc., have already taken place and are rapidly continuing. Some of these were mentioned in chapter 2 .

We feel that a summary of concepts used in the more conventional statistical theory of nuclear reactions is useful since these are expected to play a prominent role in future investigations. We proceed historically and describe the statistical model for the calculation of average cross-sections in section 8.1 , models for pre-equilibrium (or precompound) decay of the nucleus in section 8.2 , and the theoretical foundations of these models in section 8.3 . Section 8.4 gives a short account of ideas used in describing transport phenomena in deeply inelastic collisions.

8.1   The statistical model

This model was founded through the work of N. Bohr [BO 36] , Bethe [BE 37], amd Weisskopf [WE 40]. The conservation of total angular momentum was introduced into the model by Wolfenstein [WO 51] and by Hauser and Feshbach [HA 52] , and a number of refinements and theoretical investigations are due to Moldauer [MO 69] and Lane and Lynn [LA 57] . In recent years, the combination of direct and statistical processes has received considerable attention (cf. [HO 74, MO 75a] and references therein).

The statistical model is essentially based on N. Bohr's compound nucleus
picture: Once a nucleon enters a nucleus it interacts very strongly with the other
nucleons. Its mean free path is thus very short, and the energy carried by the impinging nucleon is quickly shared by many of the struck nucleons. This leads to dissipation of energy and eventually to a state of thermal equilibrium of the compound
system. By definition, the equilibrium state is independent of its mode of formation,
expect that conserved quantum numbers can, of course, not change. The particles
emitted ("evaporated") from this heated, equilibrized system will accordingly have
a distribution in energy that is typical of the total energy of the system and independent of its mode of formation ("independence of formation and decay of the
compound nucleus").

We know to-day that the mean free path of nucleons in nuclei is not as short
as would be necessary for this model to be completely correct. Modifications due to
this fact lead to the occurrence of what is called preequilibrium decay: The emission
of particles before statistical equilibrium has been reached. These modifications
are described in section 8.2 below. For simplicity of presentation, we first dis-
regard them and deal with the statistical model itself.

We simplify the problem further by assuming that direct reactions can be
neglected save, of course, for the shape-elastic scattering which is always present.
We accordingly put (cf. section 4.2)

$$\langle S_{ab}^{\mathcal{J}} \rangle = \delta_{ab} \langle S_{a\alpha}^{\mathcal{J}} \rangle . \tag{8.1.1}$$

We assume that $\langle S_{a\alpha}^{\mathcal{J}} \rangle$ is given in terms of an optical model, or one of the approximations described in chapter 5 . It is the aim of the statistical model to calculate
$\langle S_{ab}^{\mathcal{J}\, \ell} \, S_{cd}^{\mathcal{J}'\, \ell\, *} \rangle$, see section 4.2 , and to express these mean values in terms of
the $\langle S_{a\alpha}^{\mathcal{J}} \rangle$. This can be done by introducing a number of assumptions which we
justify only in section 8.3 .

We define the "transmission coefficients"

$$T_a^{\mathcal{J}} = 1 - |<S_{aa}^{\mathcal{J}}>|^2 \ , \quad 0 \leqslant T_a^{\mathcal{J}} \leqslant 1 \ . \qquad (8.1.2)$$

The coefficients $T_a^{\mathcal{J}}$ measure the unitarity deficit of the average S-matrix. Following the argument of Friedman and Weisskopf, see chapter 4.1 $\begin{bmatrix} FR \ 55 \end{bmatrix}$, we can interpret $T_a^{\mathcal{J}}$ as the probability of compound-nucleus formation from channel a . It is for this reason that in the old literature $T_a^{\mathcal{J}}$ carried the name "sticking probability". Since $<S_{aa}^{\mathcal{J}}>$ is known in terms of an optical-model description of elastic scattering in the channel a , the $T_a^{\mathcal{J}}$ are known quantities.

In deriving a value for $<S_{ab}^{\mathcal{J},\pi \ell} \ S_{cd}^{\mathcal{J}',\pi' \ell'*}>$, one first argues that different values $\mathcal{J} \neq \mathcal{J}'$ and/or $\pi \neq \pi'$ of spin and parity of the compound nucleus lead to uncorrelated fluctuations, because J and π are good quantum numbers. This means that

$$<S_{ab}^{\mathcal{J},\pi \ \ell} \ S_{cd}^{\mathcal{J}',\pi' \ \ell' *}> = \delta_{\mathcal{J}\mathcal{J}'} \delta_{\pi\pi'} <S_{ab}^{\mathcal{J},\pi \ \ell} \ S_{cd}^{\mathcal{J},\pi \ \ell *}> \ . \qquad (8.1.3)$$

We shall see that eq. (8.1.3) holds only in the frame of the statistical model, i.e. once equilibrium has been reached. It fails for the preequilibrium (or precompound) part of the reaction during which the system "remembers" the initial condition. Eq. (8.1.3) implies that $\left(d\sigma_{\alpha\alpha'}^{\ell}/d\Omega \right)$ is symmetric about 90° c.m.: The contributions of different J and π values are added incoherently, and each (J,π)-value separately leads to a symmetric angular distribution of the emitted particles. This can be verified explicitly by substituting eq. (8.1.3) into eq. (4.2.6). Experimentally observed deviations from the symmetry about 90° are thus an indication that the statistical model alone cannot explain the data, and has to be modified. As a next step one employs the assumption of complete randomness of the elements $S_{ab}^{\mathcal{J}\pi\ell}$ and finds that the expressions (8.1.3) differ from zero only if a = a' and b = b'. Hence,

$$<S_{ab}^{\mathcal{J}\pi \ \ell} \ S_{a'b'}^{\mathcal{J}'\pi' \ \ell' *}> = \delta_{\mathcal{J}\mathcal{J}'} \delta_{\pi\pi'} \delta_{aa'} \delta_{bb'} <|S_{ab}^{\mathcal{J}\pi\ell}|^2> \ . \qquad (8.1.4)$$

For a justification of this assumption and of eq. (8.1.4), c.f. section 8.3 . The expression $\langle | S_{ab}^{J\pi\ell} |^2 \rangle$ is now determined from the Bohr assumption of independence of formation and decay of the compound nucleus, formulated mathematically as

$$\langle | S_{ab}^{J\pi\ell} |^2 \rangle = \zeta_a^{J\pi} \zeta_b^{J\pi} \, . \tag{8.1.5}$$

The coefficients $\zeta_a^{J\pi}$ can be calculated from the unitarity of $S^{J\pi}$ which reads $\sum_b | S_{ab}^{J\pi} |^2 = 1$ or $\sum_b \langle | S_{ab}^{J\pi} |^2 \rangle = 1$. Hence, with eqs. (4.2.3) and (8.1.4),

$$T_a = 1 - |\langle S_{aa} \rangle|^2 = \sum_b \langle | S_{ab}^{\ell} |^2 \rangle = \zeta_a \sum_b \zeta_b \, . \tag{8.1.6}$$

For simplicity of notation, we have omitted the label (J,π). Eqs. (8.1.6) are valid for each pair (J,π). Summing eqs. (8.1.6) over a , we find $\left(\sum_b \zeta_b \right) = \sqrt{\sum_b T_b}$ and, substituting this back into eq. (8.1.6), $\zeta_a = T_a / \sqrt{\sum_b T_b}$. We insert this into eq. (8.1.5) and obtain the result

$$\langle | S_{ab}^{J\pi\ell} |^2 \rangle = \frac{T_a^{J\pi} T_b^{J\pi}}{\sum_c T_c^{J\pi}} \, . \tag{8.1.7}$$

This is the central result of the statistical model. It is usually referred to as the Hauser-Feshbach formula [HA 52] . By substituting eqs. (8.1.4) and (8.1.7) into eq. (4.2.6) we can express $d\sigma_{\alpha\alpha'}^{\ell}/d\Omega$ in terms of the transmission coefficients, i.e. in terms of the optical model of elastic scattering in anyone channel.

The result (8.1.7) is applied to two different experimental situations. One may be interested in the compound-nucleus contribution to a specific elastic or inelastic reaction. Because of the energy resolution needed, the residual nucleus must then remain in a fairly low-lying state. Alternatively, one is interested in the energy and angular distribution of the emitted particles for fixed incident energy. In the first case, the cross-section depends upon the transmission coefficients in the entrance and exit channels, and upon the sum of all transmission coefficients which occurs in the denominator of eq. (8.1.7). Since the optical-model potential is known only in the elastic channel, the transmission coefficients in the inelastic channels are unknown. They are usually computed by assuming the optical-model potentials for scattering of particles on nuclei in excited states to be the same as in ground states. To calculate the sum of transmission coefficients, one has to have a good idea of the density of states in the residual nucleus. This introduces a considerable uncertainty as a consequence of which the calculation of relative cross-sections can be performed with higher accuracy than that of absolute cross-sections. Even the former are sensitive to the optical-model potential. The angular distribution is determined by the sum over J, i.e. by the density of states of spin J in the residual nucleus. The evaluation of the sum $\sum_b T_b^{J\pi}$ can be simplified with the help of the formulas given in [EB 69]. Generally speaking, the distribution tends to be isotropic if many J-values contribute. Deviations from isotropy tend to produce a minimum at 90° c.m.. Details pertinent to such calculations can be found in [ER 66] and [VO 66] . If one is interested in the energy distribution of the emitted particles for fixed incident energy, one observes that the transmission coefficients depend essentially on the angular momentum and Coulomb barrier, and that aside from this dependence the cross-section for scattering into a group of final states is essentially determined by phase-space factors which lead to an "evaporation spectrum" of the emitted particles. This spectrum is determined by the density of states of the residual nucleus. An example of such a spectrum, obtained from proton bombardment of ^{64}Zn with particles of about 10 MeV, is shown in fig. 8.1.1 .

The comparison between the statistical model and experimental data shows, generally speaking and for reactions involving sufficiently low excitation energies of the compound nucleus, satisfactory agreement. Nonetheless, a direct test of the basic assumption (8.1.5) is very difficult. One would have to compare cross-sections of reactions induced by different projectiles, leading to the same compound nucleus at the same excitation energy. The reaction products should have the same distribution in angle and energy for both projectiles. However, different projectiles carry different angular momenta into the compound nucleus and, hence, do not lead to the same equilibrium state. A comparison of the reaction products for both projectiles, therefore, involves a theoretical estimate of the difference in distributions of angular

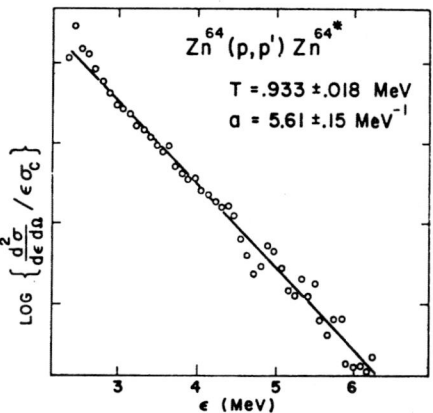

Fig. 8.1.1 Typical evaporation spectrum for protons, where T is the
nuclear temperature and a the level density parameter,
see section 3.7 . (From [JO 67]).

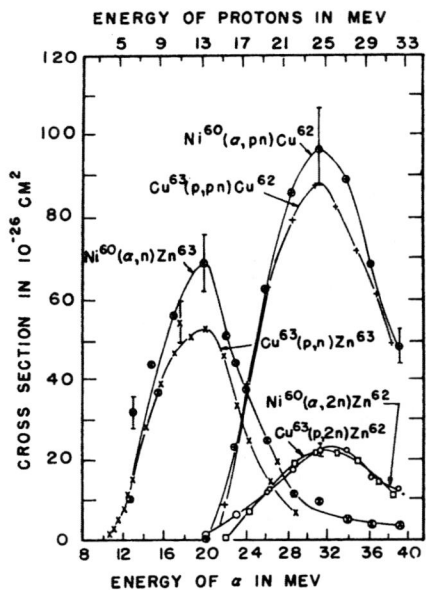

Fig. 8.1.2 Tests of the independence hypothesis. (From [GH 50]).

momenta and its consequences on the reaction. Such estimates produce an uncertainty in the analysis. Within these limitations, several such comparisons have substantiated the compound-nucleus hypothesis. An example is shown in fig. 8.1.2 .

In eq. (8.1.1) we have assumed that direct reactions are absent. In proton-induced reactions on isobaric analogue resonances, in reactions on light nuclei induced by light particles, and probably in many heavy-ion reactions, one often encounters an interplay between direct and compound processes. An extension of the statistical model to such processes is, therefore, called for. This problem has received considerable attention lately, starting with the work of Kawai, Kerman and McVoy [KA 73] . We sketch here the solution derived in [HO 74] .

The definition (8.1.2) of the transmission coefficients T_a^J has to be generalized. We introduce Satchler's transmission matrix

$$P_{ab}^{J\pi} = \delta_{ab} - \sum_c < S_{ac}^{J\pi} >< S_{bc}^{J\pi *} > . \qquad (8.1.8)$$

This matrix measures the unitarity deficit of the average S-matrix and, therefore, is connected with the compound-nucleus formation. Since $P^{J\pi}$ is Hermitean, it can be diagonalized by a unitary transformation, so that in matrix notation

$$(\mathcal{U}^{J\pi} P^{J\pi} \mathcal{U}^{J\pi\dagger})_{ab} = \delta_{ab} p_a^{J\pi} , \quad 0 \leq p_a^{J\pi} \leq 1 . \qquad (8.1.9)$$

The coefficients $p_a^{J\pi}$ are the "generalized transmission coefficients". In the absence of direct reactions, we have $\mathcal{U}^{J\pi} = 1$ and $p_a^{J\pi} = T_a^{J\pi}$. We now define the matrix

$$< \tilde{S}_{ab}^{J\pi} > = (\mathcal{U}^{J\pi} < S^{J\pi} > (\mathcal{U}^{J\pi})^T)_{ab} , \qquad (8.1.10)$$

where the superscript T denotes the transpose. It is easy to show that $< \tilde{S}_{ab}^{J\pi} >$ is diagonal with $p_a^{J\pi} = 1 - |< \tilde{S}_{aa}^{J\pi} >|^2$. This suggests the introduction of the unitary matrix

$$\widetilde{S}_{ab}^{J\pi} = \left(U^{J\pi} S^{J\pi} (U^{J\pi})^{T} \right)_{ab} \tag{8.1.11}$$

and, in view of $\langle \widetilde{S}_{ab}^{J\pi} \rangle = \delta_{ab} \langle \widetilde{S}_{\alpha\alpha}^{J\pi} \rangle$ raises the question whether the calculation of fluctuation cross-sections in terms of $\widetilde{S}_{ab}^{J\pi \, fl}$ proceeds in a manner analogous to the one described in the absence of direct reactions. This would mean that the unitary transformation introduced in eq. (8.1.9) completely transforms away the effect of direct reactions, and that the transformed matrix $\widetilde{S}$ has the same statistical properties as the S-matrix used in the statistical model without direct reactions, eqs. (8.1.3) to (8.1.7). It has been shown that the answer to this question can be given in the affirmative, so that the calculation of $d\sigma_{\alpha\alpha'}^{fl}/d\Omega$ in the presence of direct reactions is possible with the help of the unitary transformation U, and of formulas of the type (8.1.7). Details may be found in [HO 74] .

As a consequence of this result, we mention the following facts. (i) If two channels with the same (J,π), but different a,b are strongly coupled by a direct reaction, the ratio $\dfrac{d\sigma_{ab}^{fl}}{d\Omega} \Big/ \dfrac{d\sigma_{\alpha\alpha}^{fl}}{d\Omega}$ differs significantly from the prediction of the statistical model without direct reactions. (ii) The statistical model without direct reactions predicts the value zero for the compound-nucleus contribution to polarizations, particle-gamma angular correlation, etc., i.e., for all observables which cannot be written as squares of S-matrix elements. This is not true if direct reactions are included. In this case the statement just made still applies in the transformed system and for all expressions bilinear in $\widetilde{S}_{ab}^{J\pi}$. By transforming back with U to the original system, one finds, however, that the above-mentioned observables do not have vanishing expectation values. This can be expressed by saying that direct reactions induce correlations among different elements of $S_{ab}^{J\pi \, fl}$ $(J,\pi$ fixed). Such correlations have experimentally been observed both in the domain of isolated resonances (see the contributions by Lane and Mughabghab in [GA 72]), and for overlapping resonances [AB 70, GR 74, BL 75a, DA 75] .

8.2   Precompound decay

The preceding section was restricted to reactions which can be decomposed into two components, the direct and the compound part. As explained in chapter 4 and described through the models of chapter 7 , the direct part involves only few degrees of freedom. The compound reaction proceeds according to N. Bohr's hypothesis through the stage of a completely equilibrized nucleus (the compound nucleus) which is formed by the absorption of the projectile. The corresponding compound cross-section can be calculated by the Hauser-Feshbach formula (8.1.7). The angular distribution is symmetric about $90°$ and the spectrum of evaporated particles is essentially determined by the density of states of the residual nucleus. In the following we shall be concerned with deviations from this idealized compound-nucleus picture which can be attributed to particles which are emitted before the equilibrium is established. These reactions are referred to as pre-equilibrium or precompound decay. They contribute to $< S_{ab}^{J\pi \mu} \; S_{cd}^{J\pi \mu} >$ in a statistical fashion different from the compound-nucleus picture.

A schematic picture of all three (direct, precompound and compound) components in the particle spectrum (e.g. the spectrum of emitted protons in an (α, p) reaction) is shown in fig. 8.2.1 . The low energy end is essentially due to the evaporation of protons whereas the upper end shows the typical direct contribution leading to distinct final states in the residual nucleus. The contribution in between is attributed to the precompound decay. Whereas the compound contribution is symmetric about $90°$, the precompound cross-section is characterized by a forward peaking which is significant although usually less pronounced than the forward peaking of the direct reaction cross-section.

We consider the reaction mechanism in some detail as illustrated in fig. 8.2.2. The incident particle (for simplicity we assume a nucleon and neglect the Coulomb barrier for protons) strikes a target nucleon in the first step of the reaction. If the energy E of the incident particle is not large compared to the binding energy B of a nucleon, then it is likely that the incident particle is captured in a bound single-particle state without exciting a nucleon into the continuum. But at high

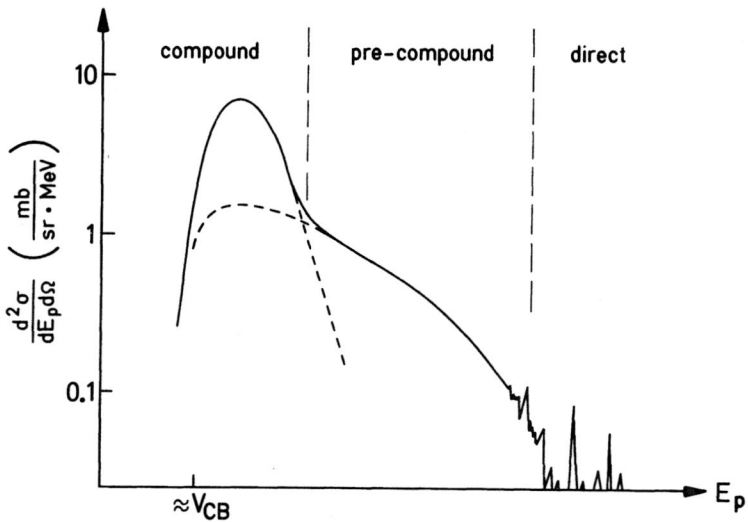

Fig. 8.2.1 Typical (schematic) spectrum of emitted protons
in an (α,p) reaction (cf. [WE 66]).

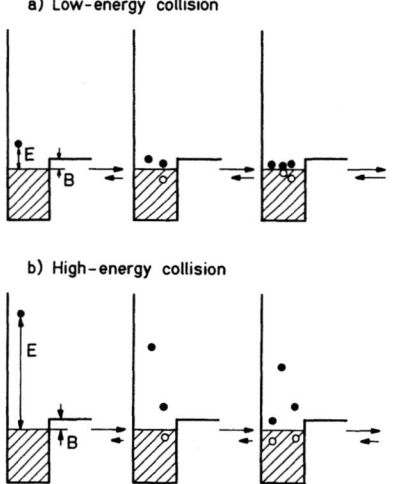

Fig. 8.2.2 Equilibration process for low-energy and high-energy
collisions.

energies E >> B   it needs many steps in order to capture all particles in bound
single-particle states. Hence, in high energy collisions the probability of emitting
particles before reaching the equilibrium state is much larger than in low energy
collisions. This is, qualitatively speaking, the reason why the statistical model
of section 8.1 ceases to work at higher excitation energies. In the following we
shall give some typical experimental results and discuss the main ideas of pre-
equilibrium models. Recent reviews on these subjects are given by Blann [BL 73a,
BL 75b] and by E. Gadioli and E. Milazzo-Colli [GA 73] .

8.2.1   Experimental results

As a typical example we discuss here results of West [WE 66] who measured
the protons which are produced by bombarding several nuclei from ^{58}Ni up to Pt with
42 MeV alpha particles. The energy spectra of the emitted protons at different
angles are very similar for all targets except Pt . In fig. 8.2.3 the spectra
from α + Rh are shown. Most of the protons are emitted at rather small energies
between 5 and 8 MeV. The angular distribution for the lower end, shown in the upper
inset, is symmetric about 90° and rather isotropic. The cross-section decreases
towards larger energies by one to three orders of magnitude and becomes strongly
forward peaked with increasing energy as shown in the lower inset. The spectra from
α + Pt , cf. fig. 8.2.4, are somewhat different, the difference being that the
maximum is at 10 MeV because of a larger Coulomb barrier and that there is already
a considerable forward peaking for the low-energy end.

8.2.2   Pre-equilibrium models

For a detailed discussion of pre-equilibrium models we refer to the review by
Blann [BL 73a, 75b] . We can distinguish two types of approaches, the intranuclear
cascade models, cf. [HA 71] , and the exciton model [GR 66, 67] which has been
combined with the master equation approach of Harp, Miller and Berne [HA 68] by
Cline and Blann [CL 71] and generalized by Blann [BL 71, 73b] . The cascade model
considers successive two-body collisions between the nucleons. The paths of the
nucleons are followed on classical trajectories in the nucleus until the nucleons

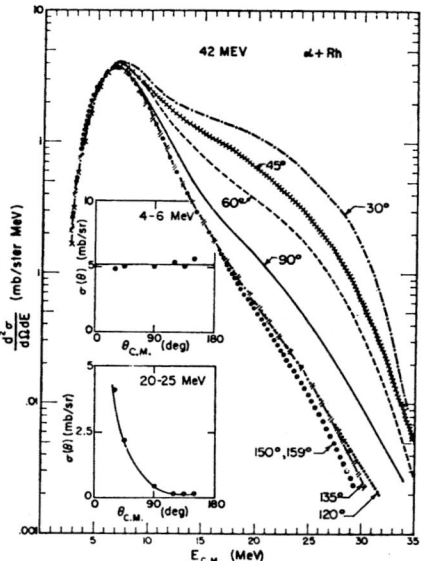

Fig. 8.2.3 Proton spectra and angular distributions for
α (42 MeV) + Rh. (From [WE 66]).

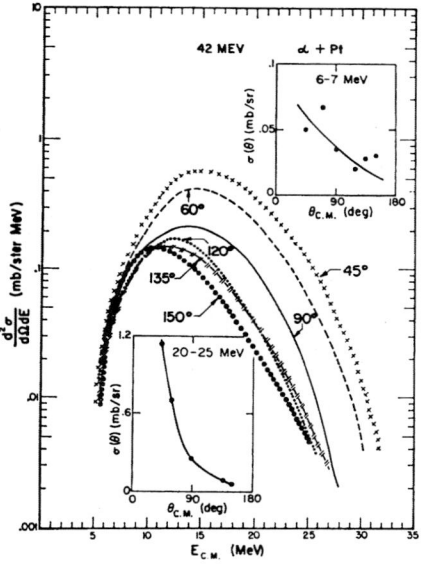

Fig. 8.2.4 Proton spectra and angular distributions from
α (42 MeV) + Pt. (From [WE 66]).

have attained energies below some given energy. The cascade model also predicts an angular distribution for the emitted particles, but the agreement of these predicted angular distributions with experimental data is rather poor for excitation energies around 50 MeV. Reasonable results can be expected from this classical model probably only for very high energies. In the second approach, one tries to follow the equilibration process in a quantum statistical formulation. Since we do not attempt to give a complete survey of the different methods which have been introduced so far, we essentially restrict the following discussion to the master equation approach of the exciton model [CL 71] which gives a nice physical picture of the equilibration process. We first discuss the calculation of angle-integrated cross-sections.

The master equation

$$\frac{d}{dt} P_n(t) = \sum_m w_{mn} \left[\rho_n P_m(t) - \rho_m P_n(t) \right] \qquad (8.2.1)$$

which is originally due to Pauli [PA 28], describes the time variation of the occupation probability $P_n(t)$ per unit energy interval of a set of states with density ρ_n. The function $P_n(t)$ increases with time by the gain $\sum_m w_{mn} \rho_n P_m(t)$ due to the decay into the considered set n of states and decreases by the loss $\sum_m w_{mn} \rho_m P_n(t)$ which is due to the decay from the set n into other states. (The quantities $w_{mn} = w_{nm}$ are mean values of squared transition matrix elements.) Hence the master equation expresses the balance between gain and loss of occupation probability.

The eq. (8.2.1) involves occupation <u>probabilities</u> (rather than amplitudes, which are determined from the Schrödinger equation). It was originally used in quantum statistical mechanics to describe the approach of macroscopic systems towards statistical equilibrium, and is the analog of the Boltzmann equation in classical statistical mechanics. Its use in nuclear physics is justified if phase correlations between different occupation amplitudes are destroyed within a short time-interval which is called the response time or memory time τ_{resp}. This time has to be short in comparison with the relaxation time or equilibration time τ_{eq} which is the time it takes the solutions of eq. (8.2.1) to reach their equilibrium values. The latter are given by the condition $\left(dP_n/dt\right)_{\tau_{eq}} = 0$ which according to eq. (8.2.1) implies

$P_n(\tau_{eq}) = \varrho_n / (\sum_m \varrho_m)$. We see that in equilibrium, all states are occupied with equal weight. Since eq. (8.2.1) works only for times $t \gg \tau_{resp}$, it cannot describe the first or first few collisions depicted in fig. 8.2.2 . Therefore, it is usually supplemented by an initial condition for $P_n(o)$ at time t = 0.

Contributions to $\langle \int d\Omega \, d\sigma^{\mu}/d\Omega \rangle$ due to precompound decay are essentially calculated as follows. The quantity $\int_0^{\tau_{eq}} P_n(t')dt'$ is a measure of the occupation probability of the states in set n during the equilibration process. Since the equilibrium configuration is reached only asymptotically, one uses in practical calculations the condition that $P_n(t)$ should differ by less than 10 per cent from the equilibrium value $\varrho_n /(\sum_m \varrho_m)$ to define the equilibration time τ_{eq} . The intensity $I(\varepsilon_c)$ of particles emitted from time zero till $t = \tau_{eq}$ into the channel c with asymptotic kinetic energy lying in the range $\varepsilon_c \ldots \varepsilon_c + d\varepsilon_c$ can be expressed by

$$I(\varepsilon_c) \, d\varepsilon_c \propto \sum_n \Gamma_{n \to c, \varepsilon_c} \, d\varepsilon_c \int_0^{\tau_{eq}} P_n(t') \, dt' \qquad (8.2.2)$$

where $\Gamma_{n \to c, \varepsilon_c}$ is the corresponding decay width of the bound state n .

The solutions $P_n(t)$ appearing in eq. (8.2.2) are determined by eq. (8.2.1) which does not take account of the decay into open channels so that eq.(8.2.2) corresponds to a first-order treatment of decay processes. Such a treatment is consistent only if the decay time $\tau_{decay} = \hbar /\langle \Gamma_{tot} \rangle$ of the system is rather large compared to τ_{eq} . This condition is often met, and precompound contributions usually constitute only a small fraction of the total reaction cross-section in reactions induced by light projectiles. This can be seen from fig. 8.2.3 where the large-angle data roughly correspond to the prediction of the statistical model. If τ_{decay} becomes comparable to τ_{eq} , the transitions into open channels must explicitly be included in eq. (8.2.1) [CL 72] .

In the exciton model, illustrated in fig. 8.1.2 for an incident nucleon, one has to consider the transitions from the subset of $[(\nu+1)p, \mu h]$ states to the neighbouring $[\nu p, (\mu-1)h]$ and the $[(\nu+2)p, (\mu+1)h]$ states where p stands for particle, h for hole. The transition probabilities are essentially given by the density ρ_n of final states, which are taken to be the usual expressions given by the Fermi gas model. A common strength factor for the transition probabilities $w_{mn} = w_{nm} = w$ is adjusted to the absolute value of the measured cross-section. (Only recently this strength factor has been calculated absolutely from the nucleon-nucleon cross-section taking the Pauli principle into account or from the mean free path of the nucleon which is obtained from the optical model [BL 73b] .) In solving the master equation, one fits the initial configuration $P_n(o)$, i.e., the initial exciton number n_i , (number of particles and holes) to the data. The numbers obtained for various cases are shown in fig. 8.2.5 . For nucleon-induced reactions this value is seen to be 3 corresponding to a (2p,1h) configuration which is expected from fig. 8.2.2 . For α-induced reactions the initial exciton number turns out to be 5 . Cline and Blann [CL 71, BL 73a] have considered the time evolution given by the master equation for a nucleus with mass ≈ 90 at 24 and 96 MeV excitation energy. The results for the occupation probabilities $P_n(t)$ as functions of the exciton number n are shown in the upper part of fig. 8.2.6 for different times in arbitrary time units depending on the (unknown) strength factor w in the transition probabilities (the time unit used is about 1% of a mean transition time). The distributions with respect to the exciton number are shown after 10, 50, 200, 800 and 3200 time units for 24 MeV excitation energy. The distribution at 3200 units of time is indistinguishable from the equilibrium distribution. For 96 MeV this equilibrium distribution is seen to be reached after 830 units of time. (Note that the time units are different for 24 and 96 MeV). The spectra calculated as functions of time t from eq. (8.2.2) with τ_{eq} replaced by t , are shown in the lower part of fig. 8.2.6 . It is seen that the high energy component is produced in the early stages of the equilibration process. As is seen from fig. 8.2.7, the pre-compound spectra $(t = \tau_{eq})$ strongly depend on the initial exciton number. Results for the calculation of pre-compound and compound spectra [CL 71] are compared in fig.8.2.8 (absolute values adjusted) with angle-averaged experimental data from [WE 66] . Absolute calculations performed recently by Blann [BL 73b] use transition rates which are obtained from the NN interaction and from the optical model. The calcu-

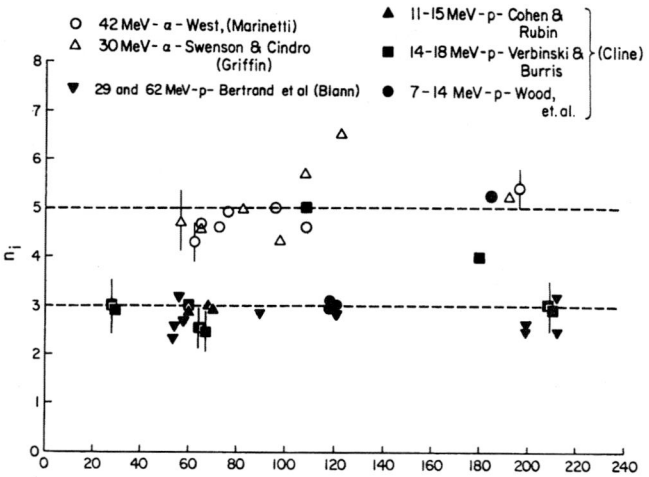

Fig. 8.2.5 Initial exciton number n_i, deduced from particle spectra
for nucleon and α-induced reactions. (From [BL 73a]).

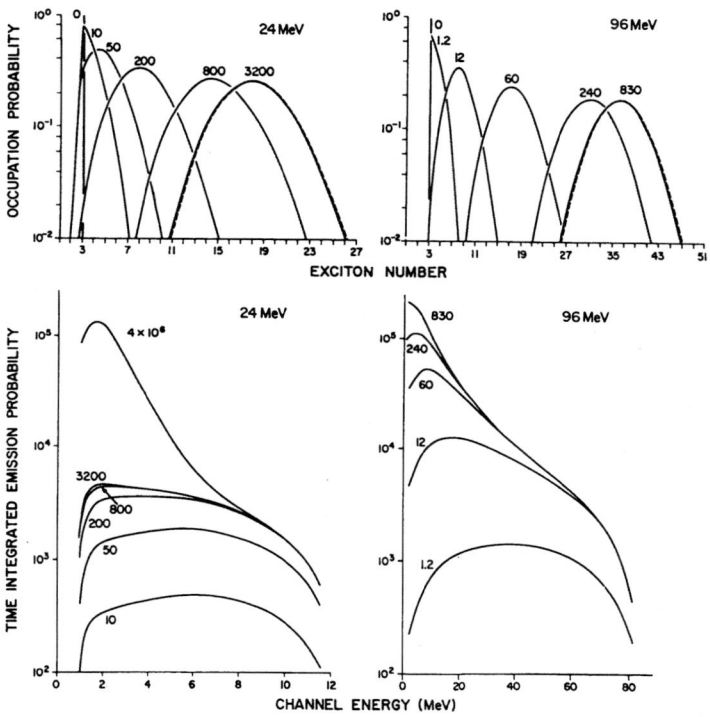

Fig. 8.2.6 Time evolution of exciton number and the time-integrated
spectra of emitted particles. (From [BL 73a]).

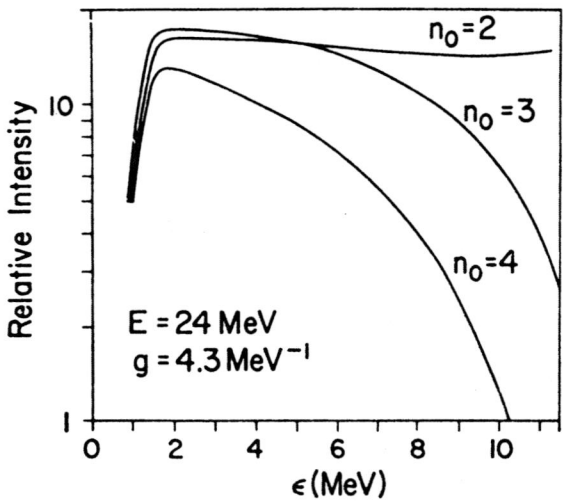

Fig. 8.2.7 Dependence of the pre-compound particle spectra on
the initial exciton number n_i. (From [CL 71]).

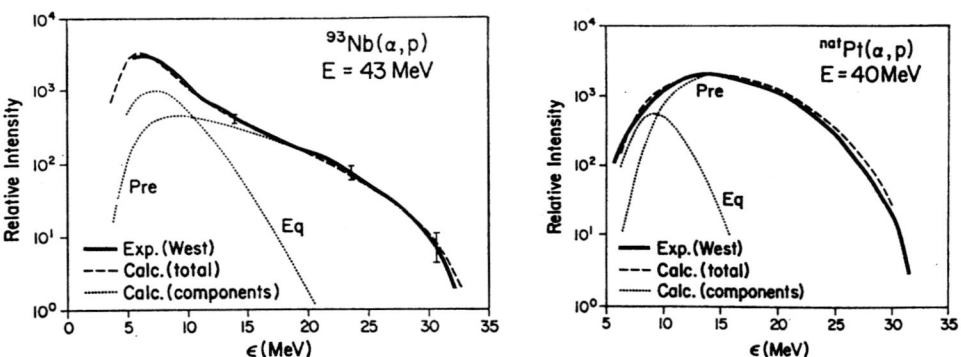

Fig. 8.2.8 Experimental and calculated spectra for α (43 MeV) + ^{93}Nb
and α (40 MeV) + Pt. (From [CL 71]).

lated cross-sections $d\sigma/d\varepsilon_c$ are off from the experimental data by only a factor 2 to 3 for $^{54}Fe(p,p')$ and $^{209}Bi(p,p')$. The resulting equilibration times turn out to be about 10^{-21} sec. They decrease with increasing energy and increase with increasing mass.

The master equation (8.2.1) can be generalized [MA 75] to give an account of the strongly forward-peaked angular distributions observed in pre-compound processes. The physical idea behind this generalization is the following. The incident nucleon, referred to as the fast particle, undergoes a series of two-body collisions with the nucleons in the target nucleus, in which it gradually loses not only its energy, but also its memory of the incident direction. This is a statistical process somewhat analogous to the energy loss of charged particles as they pass through matter and may be referred to as "angular straggling". One accordingly defines a function $P_n(t,\theta)$ as the probability of finding the system at time t in a state with n excitons and the fast particle travelling at an angle θ with respect to the incident direction. We have $\int P_n(t,\theta)\,d\Omega = P_n(t)$ where $P_n(t)$ obeys eq. (8.2.1). Eq. (8.2.1) can be generalized to include the gradual loss of memory of the incident direction; the generalized equation uses the near isotropy of the nucleon-nucleon cross-section in the c.m. system and no additional parameters that would not appear already in eq. (8.2.1) and the initial condition. Angular distributions are calculated from eq. (8.2.2) with $P_n(t)$ replaced by $P_n(t,\theta)$. Figs. 8.2.9 and 8.2.10 show that a very satisfactory account of the observed angular distributions can be given this way. The generalized form of the master equation (8.2.1) is of interest not only for the understanding of the observed angular distributions, but also because it shows how a dissipation of initial conditions other than the energy, here the incident direction, can be incorporated into a statistical treatment.

It was pointed out above that eq. (8.1.3) leads to an angular distribution which is symmetrical about 90° c.m. The deviations from this prediction and the fact that they can be described successfully imply that eqs. (8.1.3) fail to hold

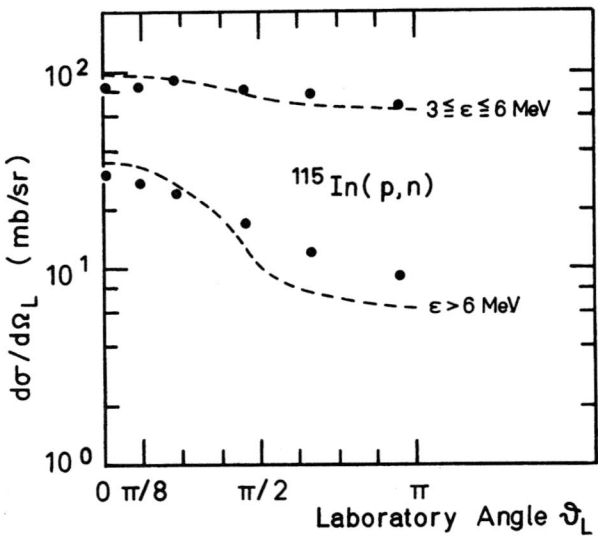

Fig. 8.2.9 The differential cross-section for the reaction ^{115}In(p,n) calculated from the model (dashed curve) in comparison with the data (points from [VE 69]), at a bombarding energy of 18 MeV. The lower part shows all neutrons with kinetic energy above 6 MeV, the upper, all neutrons with kinetic energies between 3 and 6 MeV (From [MA 75]).

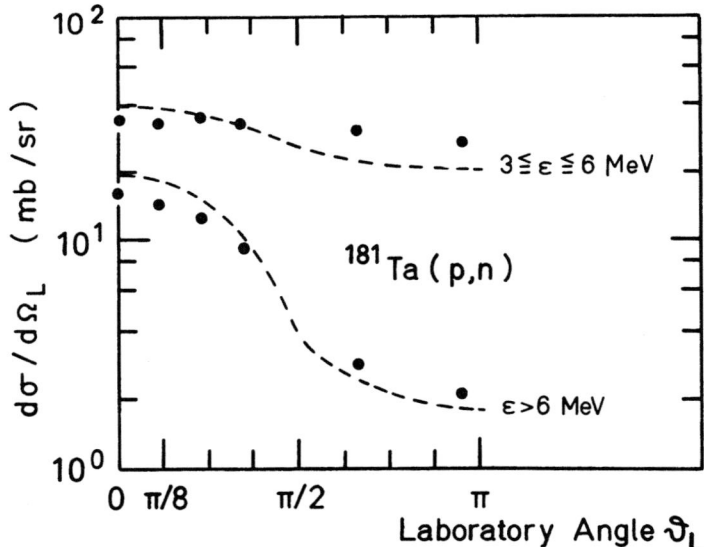

Fig. 8.2.10 The same as in fig. 8.3.9 for the reaction ^{181}Ta(p,n). (From [MA 75]).

during the pre-equilibrium phase of the reaction. This is another consequence of the fact that the system "remembers" the incident direction for a short while, i.e., until equilibrium is reached.

Although pre-compound reactions are reasonably well understood by now, there remains a number of open problems. We mention only two: (i) The description of processes lying between a direct reaction and a pre-compound one is not possible. Such reactions involve many degrees of freedom but are not sufficiently complicated to justify a statistical treatment. It is not clear at this point whether they contribute significantly to the cross-section or not. (ii) We cannot expect that level densities alone determine completely the reaction mechanism. Doorway states like giant resonances or isobaric analogue resonances are known to exist in a domain of high level density. Their influence on pre-compound processes is not known.

8.3   Theoretical foundations

In the preceding two sections, we have described theoretical approaches in which concepts are used that are familiar from quantum statistical mechanics: Statistical equilibrium, and the approach towards it. In this section we indicate how this description ties in with other statistical features of nuclear behaviour.

Statistical properties of nuclei have been observed experimentally mainly in two domains of excitation energy. In the domain of isolated resonances, mostly investigated with neutron time-of-flight spectroscopy, both the distribution of partial reduced widths and the distribution of level spacings for states of the same (J,π) show a statistical pattern. A survey of this field as well as references can be found in $\begin{bmatrix} GA\ 72 \end{bmatrix}$. In the domain of overlapping resonances, the excitation functions show statistical fluctuations, the so-called Ericson fluctuations $\begin{bmatrix} ER\ 66, \\ RI\ 74 \end{bmatrix}$. These have been shown to behave in many ways like the random noise in a radio transmitter, in keeping with the remarks made in section 4.1 above. The common theoretical basis for describing the statistical aspects of nuclear dynamics consists in the use of random matrix ensembles as introduced by Wigner and summarized by Porter $\begin{bmatrix} PO\ 65 \end{bmatrix}$. One replaces the nuclear Hamiltonian matrix by an ensemble of matrices having a statistical distribution of matrix elements and associates averages

over this (fictitious) ensemble with averages over energy of the actual physical
system, i.e., the nucleus. With the help of random-matrix theory, it is possible to
understand analytically the empirical results on the distributions of reduced width
amplitudes and on level spacings found with slow neutrons. More recently, it has
also been possible [AG 75a, AG 75b] to derive the pre-compound model of section 8.2,
the statistical model of section 8.1, and the main features of Ericson fluctuations
from random matrix theory. The pre-compound model turns out to be rather generally
valid in the domain of strongly overlapping resonances $(<\Gamma_{tot}> \gg D)$, and the
statistical model emerges if the additional condition $\tau_{decay} \gg \tau_{eq}$ is fulfilled. It
is thus seen that random matrix theory provides a common theoretical basis for statistical aspects of nuclear behaviour.

8.4   Transport theory of deeply inelastic collisions

In collisions between heavy nuclei, like Ar + Th, Cu + Au and Kr + Bi
(cf. section 2.5) the major part of the total cross-section is due to deeply inelastic processes. These deeply inelastic collisions are in many respects similar
to pre-compound reactions. In both cases the angular and energy distributions of the
outgoing particles reflect to some extent the initial momentum ("forward peaking")
and the initial energy, although these reactions are characterized by the dissipation of a large amount of momentum and energy. The pre-compound reactions are
rather well understood to-day. This is not so for deeply inelastic collisions. The
main effort in describing these processes was directed to the introduction of dissipative terms (friction forces) into the classical equations of motion (cf. chapter
2.II). Only recently, quantum-statistical methods have been introduced. For the
derivation of friction forces, Hofmann and Siemens [HO 76] have applied the
theory of linear response to the semiclassical treatment of deeply inelastic collisions (cf. section 2.7). The description of the equilibration of energy, angular
momentum and mass centers around master equations, cf. eq. (8.2.1) [NÖ 74a, NÖ 74b,
MO 75b, NÖ 75] .

By applying methods from non-equilibrium statistical mechanics and using
random matrix ensembles it is possible to derive a master equation for the equi-

libration processes which occur during the interaction between two nuclei [NÖ 74a, NÖ 75] . Because of the strong coupling between different channels, the transition probabilities w_{mn} turn out to be significantly different from those obtained by perturbation theory (golden rule). For applications in deeply inelastic collisions it is possible to transform the master equation (8.2.1) into a Fokker-Planck equation [NÖ 74b] . In the case of only one (continuous) variable x (for instance, total excitation energy or fragmentation of the nuclei) the Fokker-Planck equation for the probability distribution P(x,t) reads

$$\frac{\partial P(x,t)}{\partial t} = - \frac{\partial}{\partial x} \left[v(x,t) \, P(x,t) \right] + \frac{\partial^2}{\partial x^2} \left[D(x,t) \, P(x,t) \right] \qquad (8.4.1)$$

where the transport coefficients, i.e. the drift coefficient $v(x,t)$ and the diffusion coefficient $D(x,t)$ are completely determined by the density of channels and the second moment of the transition probability $w(x,x';t)$ with respect to the difference x-x' .

The Fokker-Planck equation is well known in statistical mechanics as a means of describing transport phenomena in gases, liquids and solids (like diffusion of particles, heat conductivity and electric current). For constant transport coefficients v and D the solution of eq. (8.4.1) is a Gaussian

$$P(x,t) = \frac{1}{\sqrt{4 \pi D t}} \; exp \left[- \frac{(x - v t)^2}{2 D t} \right] \qquad (8.4.2)$$

if initially (t = 0) the probability distribution is sharply peaked around x = 0 . As function of t the maximum of P(x,t) moves with the drift velocity v . The width is proportional to $(Dt)^{1/2}$.

The applicability of the Fokker-Planck equation for analyzing deeply inelastic collisions has been tested [NÖ 74b] in the case of the reaction $^{40}Ar + ^{232}Th$ at two different energies. Fig. 8.4.1 shows the normalized energy distributions for the element Cl and the normalized element distributions for different deflection angles. These curves are obtained from the experimental data (cf. fig. 2.5.1) by interpreting the lower energy peak as a contribution from negative deflection angles (cf. figs.2.8.7

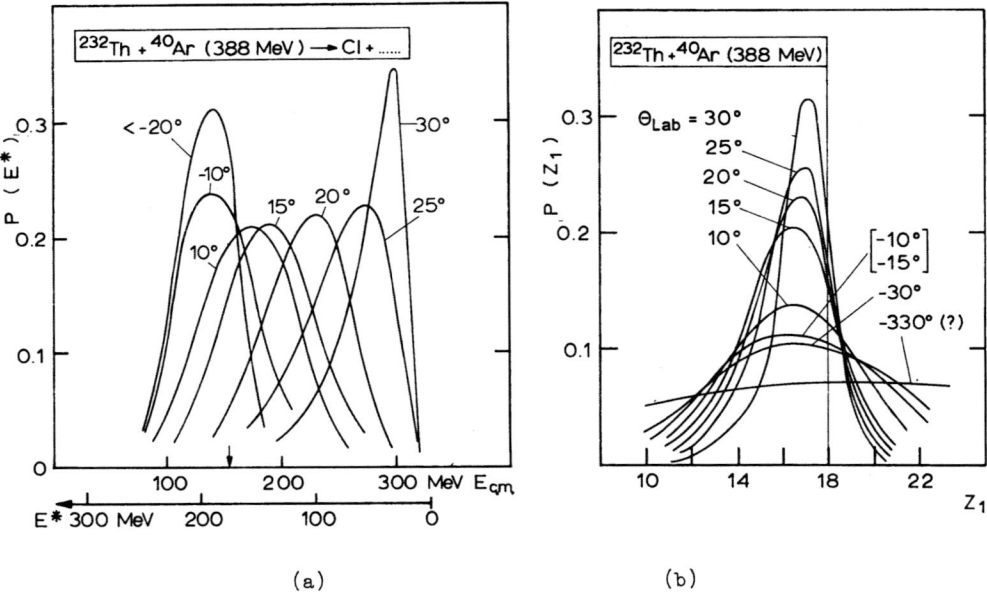

(a)   (b)

Fig. 8.4.1 (a) Normalized energy distributions of the element Cl
            and (b) normalized element distributions for different lab
            deflection angles in the collision ^{40}Ar(388 MeV) + ^{232}Th.
            (From [NÖ 74b]).

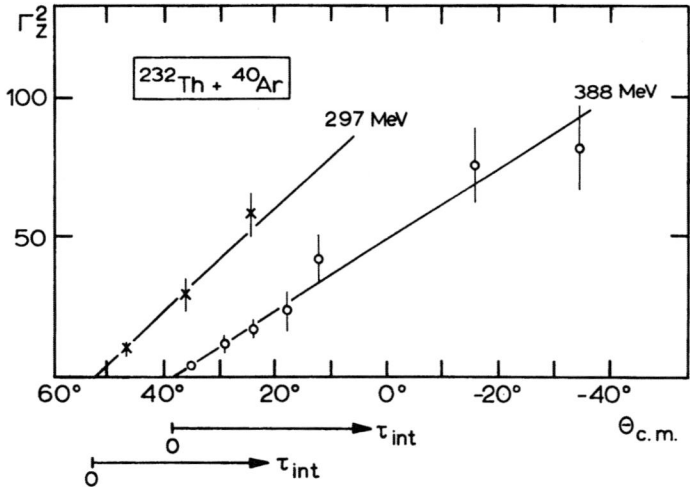

Fig. 8.4.2 Dependence of the squared width (FWHM) of the element
            distribution on the deflection angle or the interaction
            time, respectively. (From [NÖ 74b]).

and 2.8.8). The data are consistent with constant transport coefficients: Energy drift coefficient $v_E \approx 2 \cdot 10^{23}$ MeV/sec, mass drift coefficient $|v_A| \lesssim 10^{21}$ amu/sec and mass diffusion coefficient $D_A \approx 3.2 \cdot 10^{22}$ amu^2/sec if the interaction time for the deflection to $-35°$ is assumed to be $2 \cdot 10^{-21}$ sec. The significance of these transport coefficients for deeply inelastic collisions is shown also by the fact that the same values apply to the reaction Ar + Th at quite different bombarding energies (297 MeV and 388 MeV). This is illustrated in fig. 8.4.2 which shows the squared width Γ_Z^2 of the element distributions as a function of the deflection angle for both incident energies. The straight lines correspond to the same value of the diffusion coefficients D_Z if we assume that the rotational velocity of the interacting nuclei is proportional to the grazing angular momentum.

Moretto and Sventek [MO 75b] apply the master equation with a parametrized transition probability to the calculation of the equilibration in the fragmentation degree of freedom. By relating level densities occurring in the transition probability to the energy $\widetilde{E}$ of the two touching fragments (cf. section 3.2 and fig. 3.2.1) they calculate the time evolution of the composite system. This is shown in fig.8.4.3

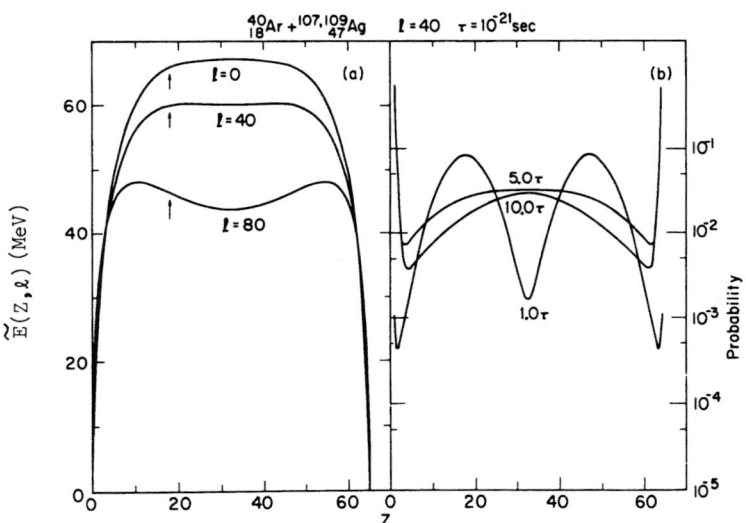

Fig. 8.4.3 (a) "Potential" energy of touching fragments as function
of the fragmentation given by the charge Z of one fragment.
(b) Time evolution of the fragmentation probability.
(From [MO 75b]).

together with the "potential" energies $\widetilde{E}(Z,\ell)$ including the rotational energy of the composite system for various angular momenta. Note that the curve for $\ell = 0$ corresponds almost to the curve with $A = 120$ of fig. 3.2.1 . The time evolution of the fragmentation is calculated for $\ell = 40$. For short interaction times $t \approx 10^{-21}$ sec the fragmentation strongly reflects the initial one. For times $t \approx 5 \cdot 10^{-21}$ sec the heavy and light peaks merge into one another. For $t \approx 10^{-20}$ sec the fragmentation distribution has practically reached the equilibrium distribution. Assuming that the rotating composite system decays with a time constant $\tau_{dec} = 2.5 \cdot 10^{-21}$ sec into two separated fragments, Moretto and Sventek can account for the main features of the angular distributions of the fragments in the collision ^{40}Ar(288 MeV) + Ag .

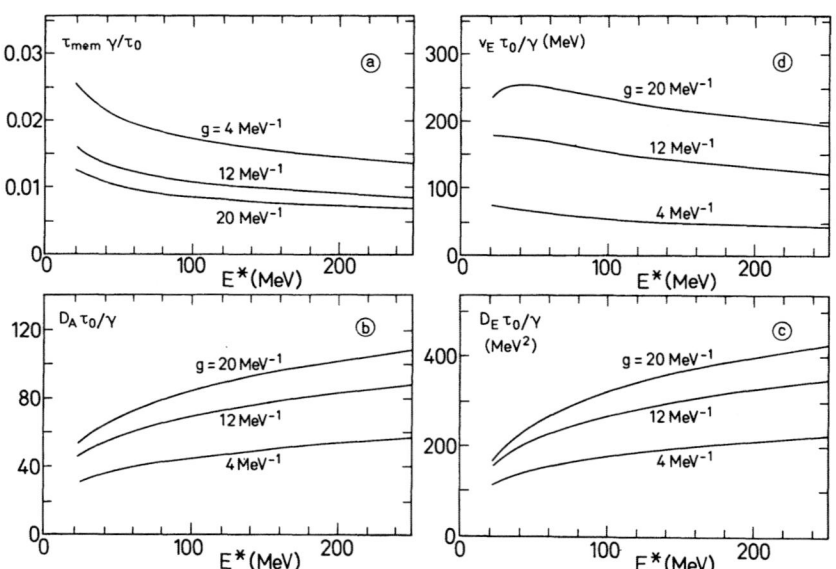

Fig. 8.4.4 Memory time τ_{mem} (E^*,A) and transport coefficients of mass diffusion $D_A(E^*,A)$, energy diffusion $D_E(E^*,A)$ and energy drift $v_A(E^*,A)$ as functions of the excitation energy E^* for different masses $A = A_1 = A_2$ of the colliding nuclei ($g = A/12$ MeV^{-1}). Here, $\tau_0 = 2\pi\hbar/1$MeV $\approx 4 \cdot 10^{-21}$ sec is used as a time unit. The dimensionless quantity γ measures the interaction strength and has according to experimental data typical values around 3 . (From [AY 75]).

First calculations of transport coefficients have been performed around symmetric fragmentations [AY 75] using the expression obtained in [NÖ 75] for the transition probabilities. The total excitation energy E^* of the fragments and the fragmentation denoted by the fragment mass number A_1 , are taken as variables. Without going into details of these calculations we note the following results for the symmetric fragmentation $A_1 = A_2 = A$ which are illustrated in fig. 8.4.4 .

(a) The memory time $\tau_{mem}(E^*, A_1 = A)$ is slightly decreasing with increasing excitation energy E^* and increasing mass A of the colliding nuclei. This memory time has to be small as compared to all other characteristic times in order that the master equation is valid. A typical value is $\tau_{mem} \approx 10^{-23}$ sec which is, indeed, sufficiently small.

(b) The mass diffusion coefficient $D_A(E^*, A_1 = A)$ increases slightly with increasing energy and mass of the composite system.

(c) The energy diffusion coefficient D_E has a similar dependence on the excitation energy and mass, but increases somewhat more steeply with excitation energy.

(d) The energy drift coefficient v_E is practically independent of the excitation energy and increases linearly with the mass.

The mass drift coefficient v_A as well as the mixed diffusion coefficient $D_{EA} = D_{AE}$ are zero for the symmetric fragmentation. In calculating these transport coefficients it is assumed that the transfer of energy per elementary step is limited to the range of ≈ 2 MeV.

In the next few years we expect strong activities in the further development of the transport theory in HI reactions along the lines sketched above. We believe that the study of this field will contribute towards our understanding of statistical features of nuclei. Moreover, these processes represent equilibration phenomena for rather small quantum systems which initially are far from equilibrium or stationarity. Therefore, they may also be considered as interesting examples for testing the validity and limitations of concepts used in non-equilibrium statistical mechanics.

References

AB 70   E. Abramson, R.A. Eisenstein, I. Plesser, Z. Vager, J.P. Wurm,
Nucl.Phys. $\underline{A144}$ (1970) 321

AG 75a   D. Agassi and H.A. Weidenmüller, Phys. Lett. $\underline{56B}$ (1975) 305

AG 75b   D. Agassi, H.A. Weidenmüller and G. Mantzouranis, Phys.Lett.$\underline{22C}$ (1975)145

AY 75   S. Ayik, B. Schürmann and W. Nörenberg, Z.Physik (in press)

BE 37   H.A. Bethe, Revs. Mod. Phys. $\underline{9}$ (1937) 71

BL 71   M. Blann, Phys. Rev. Lett. $\underline{27}$ (1971) 337

BL 73a   M. Blann, in Lecture Notes in Physics, vol. 22, ed. by N.Cindro et al.
(Springer Verlag, Berlin - Heidelberg - New York, 1973) p. 43

BL 73b   M. Blann, Nucl. Phys. $\underline{A213}$ (1973) 570

BL 75a   E. Blanke, H. Genz, A. Richter and G. Schrieder, Phys.Lett. $\underline{58B}$ (1975) 289

BL 75b   M. Blann, Ann. Rev. Nucl. Science $\underline{25}$ (1975) 123

BO 36   N. Bohr, Nature $\underline{137}$ (1936) 344

CL 71   C.K. Cline and M. Blann, Nucl. Phys. $\underline{A172}$ (1971) 225

CL 72   C.K. Cline, Nucl. Phys. $\underline{A193}$ (1972) 417

DA 75   S. Davis, C. Glashausser, A.B. Robbins, G. Bissinger, R. Albrecht, J.P.Wurm,
Phys. Rev. Letts. $\underline{34}$ (1975) 215

EB 69   K.A. Eberhard, P. von Brentano, M. Böhning and R.O. Stephen,
Nucl. Phys. $\underline{A125}$ (1969) 673

ER 58   T. Ericson and V.M. Strutinsky, Nucl. Phys. $\underline{8}$ (1958) 284 and $\underline{9}$ (1958) 689

ER 66   T. Ericson and T. Mayer-Kuckuk, Ann. Revs. Nucl. Science $\underline{16}$ (1966) 183

FR 55   F.L. Friedman and V.F. Weisskopf, in "Niels Bohr and the Development of
Physics", ed. by W. Pauli (Pergamon Press, New York, 1955) p.134

GA 72   Statistical Properties of Nuclei, ed. by J.B. Garg (Plenum Press, New York -
London, 1972)

GA 73   E. Gadioli and L. Milazzo-Colli, in Lecture Notes in Physics, vol. 22,
ed. by N. Cindro et al. (Springer Verlag, Berlin - Heidelberg - New York,
1973) p. 86

GH 50   S.N.Ghoshal, Phys. Rev. $\underline{80}$ (1950) 939

GR 66   J.J. Griffin, Phys. Rev. Lett. $\underline{17}$ (1966) 478

GR 67   J.J. Griffin, Phys. Lett. $\underline{B24}$ (1967) 5

GR 74   G. Graw, H. Clement, J.H. Feist, W. Kretschmar and P. Pröschel,
Phys. Rev. C10 (1974) 2340

HA 52   W. Hauser and H. Feshbach, Phys. Rev. 87 (1952) 366

HA 68   G.D. Harp, J.M. Miller and B.J. Berne, Phys. Rev. 165 (1968) 1166

HA 71   G.D. Harp and J.M. Miller, Phys. Rev. C3 (1971) 1847

HO 74   H.M. Hofmann, J. Richert, W. Tepel and H.A. Weidenmüller,
Ann. Phys. (N.Y.) 90 (1975) 403

HO 76   H. Hofmann and P. Siemens, Nucl. Phys. A257 (1976) 165

JO 67   R.R. Johnson and N.M. Hintz, Phys.Rev. 153 (1967) 1169

KA 73   M. Kawai, A.K. Kerman, and K.W. McVoy, Ann.Phys. (N.Y.) 75 (1973) 156

LA 57   A.M. Lane and J.E. Lynn, Proc. Phys. Soc. (London) A70 (1957) 557

MA 75   G. Mantzouranis, D. Agassi and H.A. Weidenmüller, Phys. Lett. 57B (1975)220

MO 69   P.A. Moldauer, Phys. Rev. 177 (1969) 1841 and references therein to earlier
work by the same author

MO 75a   P.A. Moldauer, Phys.Rev. C11 (1975) 426

MO 75b   L.G. Moretto and J.S. Sventek, Phys. Lett. 58B (1975) 26

NÖ 74a   W. Nörenberg, Proc.Int.Conf. on Reactions between Complex Nuclei, Nashville,
U.S.A., 1974 (North-Holland, Amsterdam, 1974) vol. 1, p. 90

NÖ 74b   W. Nörenberg, Phys. Lett. 52B (1974) 289

NÖ 75   W. Nörenberg, Z. Physik A274 (1975) 241; Erratum A276 (1976) 84

PA 28   W. Pauli, Festschrift zum 60. Geburtstag Sommerfelds, ed. by P. Debye
(Hirzel-Verlag, Leipzig, 1928) p. 30

PO 65   C.E. Porter, Statistical theories of spectra: Fluctuations (Academic Press,
Inc., New York, 1965)

RI 74   A. Richter, in Nuclear Spectroscopy and Reactions, ed. by J. Cerny,(Academic
Press, New York,1974)

VO 66   W. von Witsch, P. von Brentano, T. Mayer-Kuckuk, A. Richter, Nucl.Phys.80
(1966) 394

VE 69   V.V. Verbinsky and W.R. Burrus, Phys.Rev. 177 (1969) 1671

WE 40   V.F. Weisskopf and D.H. Ewing, Phys. Rev. 57 (1940) 472 and 935

WE 66   R.W. West, Phys. Rev. 141 (1966) 1033

WO 51   L. Wolfenstein, Phys. Rev. 82 (1951) 690

9.   Atomic effects in ion-atom collisions

In the previous chapters we have considered the effects which result from the interaction between the colliding nuclei. Nothing has been said so far about the behaviour of the electrons in the inner shells during the HI collision. We shall discuss here two of the most interesting and still unsettled problems: (i) The emission of molecular X-rays from quasiatoms which are formed during the collision. The basic experimental and theoretical work on the behaviour of inner-shell electrons in an ion-atom collision is reviewed in $\begin{bmatrix} \text{GA 73, KE 73, CR 75} \end{bmatrix}$. (ii) The spontaneous production of positrons in over-critical fields of superheavy quasi-atoms. The spontaneous production of positrons can occur whenever a vacancy in a bound electron state exists with a binding energy larger than $2m_e c^2$ (m_e = electron mass). The existence of such strongly bound states has first been studied by Pomeranchuk and Smorodinskii $\begin{bmatrix} \text{PO 45} \end{bmatrix}$ cf. $\begin{bmatrix} \text{AK 65} \end{bmatrix}$. A vacancy in a K shell which is bound by more than $2m_e c^2$ can be filled spontaneously by electron-positron pair production, the electron filling the vacancy and the positron being emitted $\begin{bmatrix} \text{VO 61, BE 63, GE 69, PI 69} \end{bmatrix}$. Such a process becomes possible in HI collisions and has been studied

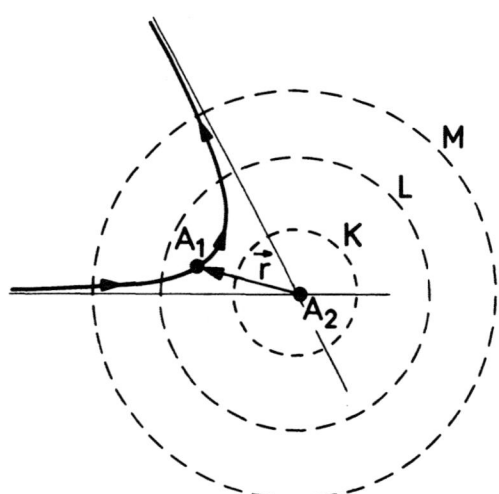

Fig. 9.1.1 A trajectory of nucleus A_1 through the electronic
             L shell of nucleus A_2

extensively by Greiner and coworkers [PI 69, RA 71, MÜ 72, 73, PE 73, SO 73, MÜ 74 SM 74, BE 75] and by Popov and coworkers [PO 70, 71a,b, MI 72, PE 72, PO 73, 74, MA 75], cf. also the review article by Zeldovich and Popov [ZE 72].

9.1 Production of inner-shell vacancies

We consider here the collision between two heavy ions at energies below the Coulomb barrier. The relative velocity of the nuclei is in general much smaller than the velocities of the electrons in the inner shells which are just penetrated by the other nucleus (cf. fig.9.1.1). Typically we have

$$v_{nucleus} / v_{electron} \lesssim 0.1 \tag{9.1.1}$$

even for the K-shell electrons of U on U with energies below the Coulomb barrier. Hence, we may expect to obtain reasonable solutions of the Schrödinger equation within the Born-Oppenheimer approximation where the wave function $\bar{\psi}$ for the system is a product of nuclear and electronic parts, i.e.

$$\bar{\psi} = \chi_N(\vec{r},t)\, \psi_e(\vec{r},\vec{r}_e) \; . \tag{9.1.2}$$

Here, $\vec{r}_e$ represents the electronic coordinates and $\vec{r}$ the distance between the nuclear centers. If $\vec{r}$ is assumed to be completely fixed, the electronic part would obey the equation

$$H_e(\vec{r},\vec{r}_e)\, \psi_e(\vec{r},\vec{r}_e) = E_e(r)\, \psi_e(\vec{r},\vec{r}_e) \tag{9.1.3}$$

where $\vec{r}$ is considered as a parameter. The Born-Oppenheimer approximation consists in using ψ_e from eq. (9.1.3) in eq. (9.1.2), χ_N being determined by

$$\left[H_N + E_e(r) \right] \chi_N(\vec{r},t) = i\hbar \frac{\partial}{\partial t} \chi_N(\vec{r},t) \; . \tag{9.1.4}$$

Here, the action of H_N on ψ_e has been neglected. The stationary states defined by eq. (9.1.3) are then considered to be perturbed by H_N, which causes electronic transitions between the states ψ_e.

Within the Born-Oppenheimer approximation each orbital of the ion-atom system has an energy which is a function of the internuclear distance r , [HU 27]. It may happen that the energies of two such orbitals with the same quantum numbers approach each other at $r = r_x$. Since these orbitals result from the diagonalization of the same Hamiltonian, they are repelled and avoid crossing. Fano and Lichten [FA 65, LI 67, BA 72, LI 74] have proposed, however, that a better understanding of collisions involving inner-shell electrons is obtained by defining a set of states which run smoothly through the crossings. Such "diabatic" states can be well defined in terms of one-electron molecular orbitals at all values of the internuclear distance. The corresponding level diagram, usually called correlation diagram, is shown in fig. 9.1.2 for the system Ar + Ar . At each crossing there is a non-vanishing probability that an electron occupying one level makes a transition to the other level. The strongest coupling arises from nuclear motion, both radial and rotational (Coriolis coupling). A much weaker coupling is due to spin-orbit and residual interaction (configuration mixing). By this mechanism, electrons from inner shells can be promoted to unfilled shells. Hence, after the collision the atoms are left in excited states with holes in their inner shells.

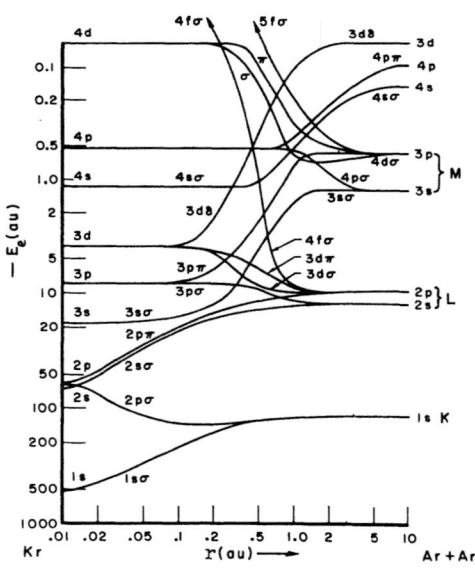

Fig. 9.1.2 Correlation diagram of Ar + Ar as presented by Fano
and Lichten. (From [FA 65] .)

In addition, holes in inner shells can be produced by the following mechanism:
(i) In symmetric collisions like Ar + Ar , K-vacancies are primarily produced by
the coupling of the 2pσ and 2pπ molecular orbitals at very small internuclear
distances. This coupling is provided, in a diabatic model (cf. fig. 9.1.2), by the
angular part of the nuclear kinetic energy operator (Coriolis coupling); i.e. by the
sudden rotation of the internuclear axis near the distance of closest approach.
(ii) In very asymmetric collisions like Ar + Cu a K-shell vacancy in the heavier
atom can be produced by neither of the two mechanisms (cf.fig.9.1.3). In such cases
holes in the K-shell of the heavier atom may be produced by direct Coulomb excitation.
However, the cross-section is small for this. process. Other mechanisms have not been
studied so far.

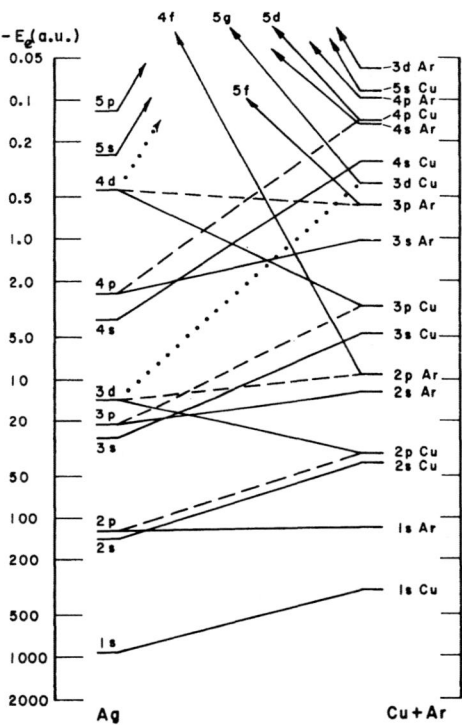

Fig. 9.1.3 Schematic correlation diagram for the Ar + Cu
system.(From [BA 72]).

9.2   Molecular X-rays from quasi-atoms

Consider now the holes which have been formed during the collision. These vacancies can be filled by X-ray emission or by emission of Auger electrons. If the X-ray emission happens after the collision when the atoms are again separated, the corresponding X-rays have energies which are characteristic for the atoms. In addition to these characteristic X-rays of the separated atoms, non-characteristic continua of X-rays have been observed (already by Coates [CO 34]) from the quasi-atoms which are formed during the collision. Saris et al. [SA 72] have observed a broad line, or band, with centroid energy of about 1 keV which they attribute to the radiative filling of the 2pπ hole of the Ar + Ar quasimolecular system during the collision. In order that the observed transition can occur during the collision, it is necessary that the projectile carries a 2p-hole into the 2pπ molecular orbit. This hole must have been formed previously, i.e. in a prior collision in the solid target. This picture has been confirmed by the failure to observe molecular X-rays in a dilute gas target, where the 2p-holes decay between two collisions. For the filling of the 2pπ vacancy during the second collision, the lifetime of the 2pπ hole must be of the same order as the duration time of the collision which is $\approx 10^{-16}$ sec for the 100 keV Ar projectile. It is known that the lifetime of a 2p vacancy in Kr is 1 order of magnitude shorter than the lifetime of a 2p vacancy in Ar which is $4 \cdot 10^{-15}$ sec. Consequently, the lifetime of a 2p hole is, indeed, expected to be comparable to the duration time of the collision.

Mokler et al. [MO 72] have observed X-rays from superheavy quasiatoms which are formed during the collisions of I ions on Au, Th and U . Fig. 9.2.1 shows these spectra together with spectra from I + Yb and Pb, [AR 74, MO 74] . In all spectra a broad X-ray peak indicated by arrows and lying at $\approx$ 7 keV for I + Yb and at $\approx$ 10 keV for I + U is observed. The width of the peaks is consistent with the collision time (collision broadening). With increasing atomic number of the quasi-atom,the position of the peak shifts to higher energies. Whereas the molecular X-rays in the collision of Ar ions on Ar have been attributed to the filling of L-shell vacancies, the broad peaks of fig. 9.2.1 are expected to be produced by the filling of M-shell vacancies. Indeed, the peaks are always found near the expected M_α and M_β transitions in the corresponding superheavy atom [FR 74] . The M transitions are indicated by the arrows in fig. 9.2.1 . A calculated correlation diagram for the

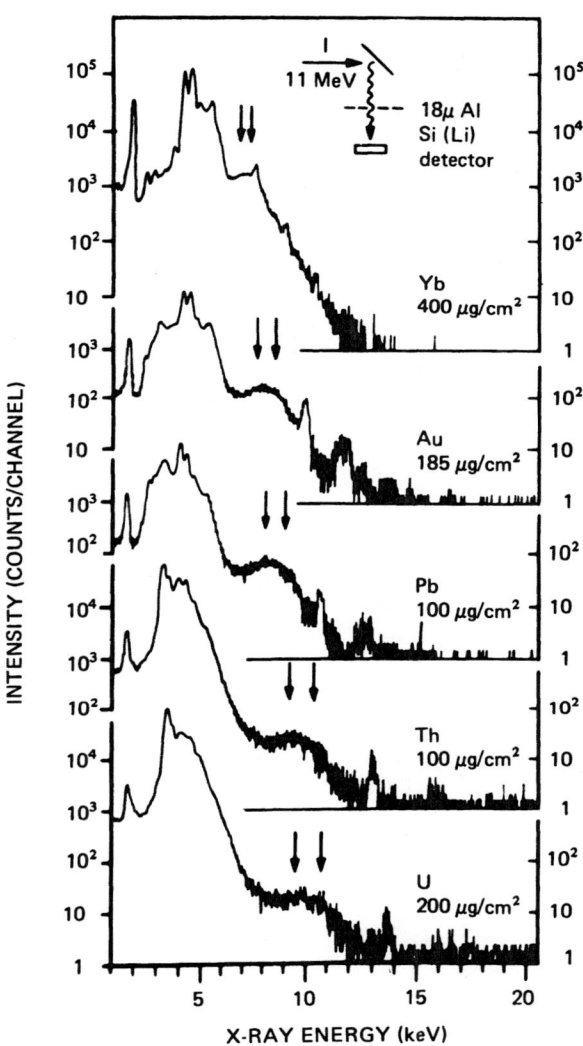

Fig. 9.2.1 X-ray spectra from the collisions I(6 MeV) + Yb and
            I(11 MeV) + Au, Pb, Th, and U obtained with an 18 μm
            Al filter. (From [MO 74]).

I + Au molecule with 20 electrons is shown in fig. 9.2.2 . The observed molecular
X-rays are attributed to the transition 4f → 3d . There are two main unsettled
problems in the interpretation of these molecular X-rays: (i) The observed aniso-
tropy of the radiation and (ii) its large cross-section. It has been suggested
[ST 73, MÜ 74, BE 75] that both effects are due to transitions which are induced
by the time-varying field of the moving nuclei, but the large cross-section cannot
be explained by this mechanism. Data taken on solid and gaseous targets show that
the double collision mechanism (as discussed for the Ar + Ar collision) is no longer
operative here. Instead of being produced in a previous collision, the inner-shell
vacancy is produced and decays during the same atomic collision [ME 73, AR 75] .

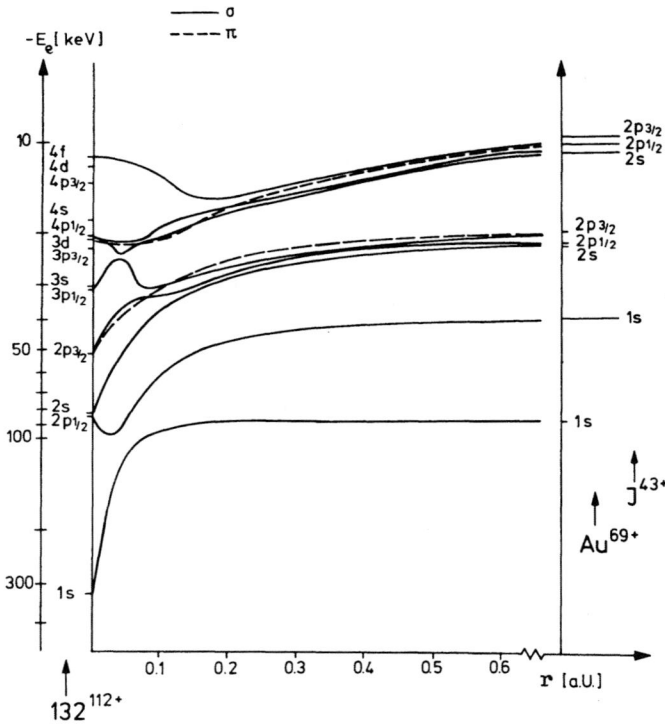

Fig. 9.2.2 Calculated correlation diagram for 20 electrons
in the I-Au molecule. (From [MO 74]).

Apart from molecular M-shell X-rays one has also studied molecular L- and K-shell X-rays in collisions between heavy atoms, cf. for example [MA 73, ME 73, DA 74, GI 74, GO 74, GR 74, ME 74, AN 75, FR 75] . In all spectra a continuum of noncharacteristic X-rays is observed. The intensity decreases with increasing energy. No sharp end point has been found. These continua have been studied recently by Anholt and Saylor [AN 75] in the bombardment of Zr by 200 MeV Kr, 60 MeV Cl and 12-35 MeV O, and in collisions of 200 MeV Kr on Ti, Ni and KCL. The observed spectra plotted per K-shell vacancy of Zr and Kr, respectively, are shown in fig. 9.2.3 . The X-ray spectra turn out to be independent of the projectile-target combination and within the observed energy range also of the velocity. At present, there is no satisfactory interpretation of these data available. The molecular orbital picture, for instance, leads us to expect that the continuous spectra for O + Zr and Kr + Zr should differ in shape and magnitude. For O + Zr the spectrum should approach an end point much lower in X-ray energy.

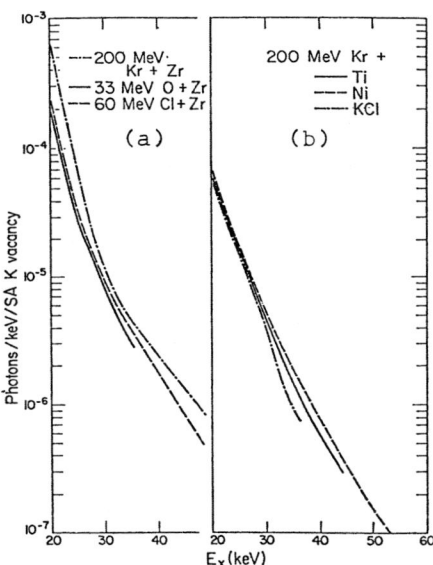

Fig. 9.2.3 (a) X-ray spectra per K-vacancy of Zr for 200 MeV Kr, 60 MeV Cl and 33 MeV O on Zr. (b) X-ray spectra per K-vacancy of Kr for 200 MeV Kr on Ti, Ni and KCl. (From [AN 75]).

9.3   Interior electron shells in super-heavy atoms

In the following two sections we discuss electronic effects which are connected with strong electromagnetic fields. Such strong electromagnetic fields are
produced in a close collision of two heavy ions if the sum of the nuclear charges
becomes considerably larger than 137 .

We consider first the electronic structure of spherical nuclei as a function
of the charge number Z (cf. fig. 9.3.1). For Z = 0, one finds two kinds of continuum
states as solutions of the free Dirac equation, the positive energy states $(E_e \geq m_e c^2)$
and the negative energy states $(E_e < -m_e c^2)$. In the vacuum (no physical electrons
and positrons) all negative energy states are occupied by electrons. A hole in this
sea is interpreted as a positron. For Z > 0 additional bound states occur below the
positive energy continuum. For a point nucleus the 1s state has the energy

$$E_{1s} = m_e c^2 \sqrt{1 - Z^2 \alpha^2} \qquad\qquad (9.3.1)$$

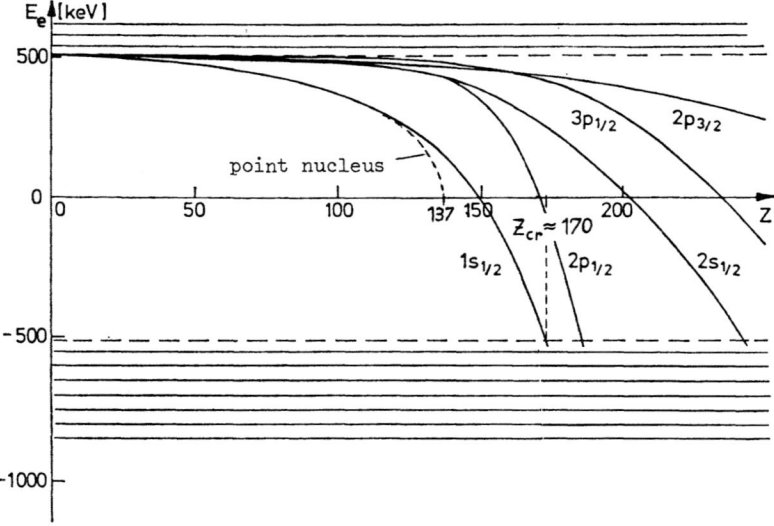

Fig. 9.3.1 Level diagram of the lowest electron levels as
          function of the nuclear charge number.
          (From [MÜ 72a]).

where $\alpha = 1/137$ is the fine-structure constant. These solutions for a point
nucleus cannot be extended to $Z > 137$. If one uses the more realistic Coulomb
potential of a homogeneously charged sphere instead of the singular $1/r$ potential,
the solutions of the Dirac equation can be obtained up to the point where the levels
join the negative energy continuum (cf. fig. 9.3.1). For the 1s state the corresponding critical Z value is $Z_{cr} \approx 170$ [VO 61, PI 69, RE 69, PO 70, 71a, 71b, MÜ 72a,
MA 75].

There exist corrections to Z_{cr} due to vacuum polarization and due to the
fact that in a heavy-ion scattering we have two centers instead of one. The change
in Z_{cr} due to many-body effects (vacuum polarization) is expected [ZE 72] to be
$\approx \alpha \, Z_{cr} \approx 1$. Moreover, the study [RA 71, SO 73] of models of maximal vacuum
polarization (limiting field theories) has shown that the vacuum polarization cannot
prevent the diving of bound states into the negative continuum. The presence of two
centers leads to a strong quadrupole interaction. This can shift $Z_{cr} = (Z_1 + Z_2)_{cr}$
also only by a few units [MÜ 73].

9.4   Production of positrons in overcritical fields

We have seen that the 1s 1/2 state reaches the lower continuum at a critical
nuclear charge $Z_{cr} \approx 170$. The calculations can be carried beyond the point Z_{cr}
by a model [MÜ 72a,b; similar results have been obtained in VO 61, MI 72, PE 72
ZE 72], originally due to Dirac [DI 47], introduced by Fano [FA 61] to
atomic scattering problems and developed in nuclear reaction theory [MA 69].

Following Dirac [DI 47] we assume that the solutions of the Dirac equation
are known for $Z = Z_{cr}$. Let $|\phi\rangle$ denote the bound-state wave function of the 1s
state, and $|\psi_E\rangle$ the negative energy continuum states. Let $V(r_e, Z') = Z'U(r_e)$
denote the change in Coulomb potential due to changing Z from Z_{cr} to larger
values $Z = Z_{cr} + Z'$. An approximate solution $|\chi_E\rangle$ for $Z = Z_{cr} + Z'$ is
obtained by expanding $|\chi_E\rangle$ in terms of the solution obtained at $Z = Z_{cr}$,

$$|\chi_E\rangle = a(E)|\phi\rangle + \int dE' \, b_{E'}(E) |\psi_{E'}\rangle \quad . \tag{9.4.1}$$

Following Fano, we assume that $\langle \psi_E, |V| \psi_E \rangle$ is negligibly small, and that the dependence of $\langle \psi_E |V| \phi \rangle$ on energy can be neglected. Then, a straightforward calculation yields [MÜ 72a,b]

$$| a(E) |^2 \approx \frac{1}{2\pi} \frac{Z'^2 \gamma}{[(E + m_e c^2) + Z'\delta]^2 + [Z'^2 \gamma / 2]^2} \qquad (9.4.2)$$

where $\gamma = 2\pi |\langle \psi_E |U| \phi \rangle|$ and $\delta = -\langle \phi |U| \phi \rangle$. The amplitude $a(E)$ displays a maximum of Breit-Wigner shape at energies below $-m_e c^2$. This means that the bound state $|\phi\rangle$ dives into the negative energy continuum for $Z > Z_{cr}$, the diving depth being proportional to $Z' = Z - Z_{cr}$. Simultaneously, it is spread over the neighbouring continuum states, the width being given by $Z'^2 \gamma$. This behaviour is illustrated in fig. 9.4.1. Typical values for γ and δ are 0.05 keV and 30 keV, respectively. The simple Breit-Wigner form is expected to be modified by penetration effects. The physical meaning of the diving into the continuum and the occurrence of

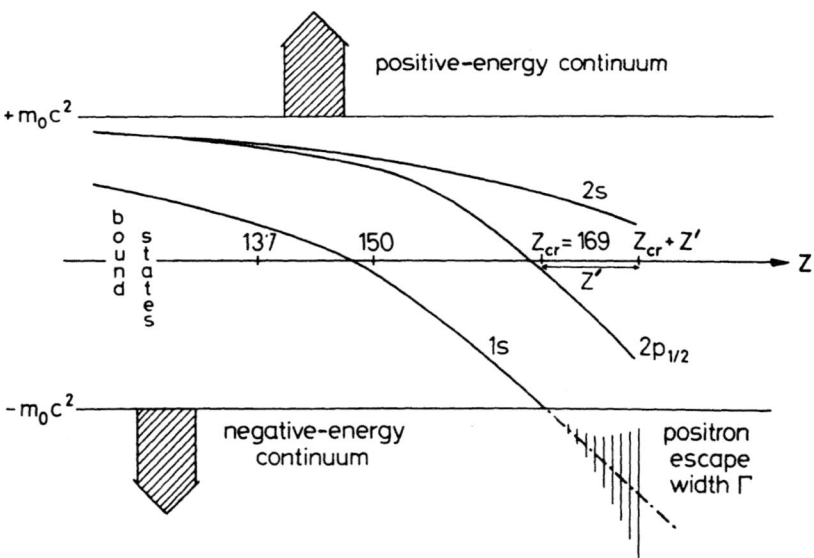

Fig. 9.4.1 The lowest atomic levels as function of the nuclear charge. The diving of the bound 1s state into the continuum and the dependence of the width Γ is illustrated. (From [MÜ 72b]).

a decay width is illustrated in fig. 9.4.2 . For $E \geqslant m_e c^2$ there is the normal
electron continuum with wave functions (indicated by wavy lines) extended over the
whole r_e-space. Below $m_e c^2$ there are bound electron states. The positron states
are indicated in this diagram by wavy lines below $-m_e c^2$. Because of the Coulomb repulsion, there is a gap in the r_e-space between the bound 1s-state and the negative
continuum states. By the coupling due to $Z'U(r_e)$, the bound 1s-state is spread
over the continuum states. This means that if initially a hole is produced in the
bound 1s-state, the hole decays (decay or escape width Γ) giving rise to a positron
in the continuum. This process can be considered as a pair production [BE 63] where
the electron is created in the bound 1s-state and the positron in a continuum state.

We consider now a rough estimate [PE 73] of the cross-section for this
spontaneous production of positrons in a heavy-ion collision like U + U . In order
to describe the diving of the 1s-state in a simple way as function of r , we define
an effective charge number $Z(r)$ and use the escape width

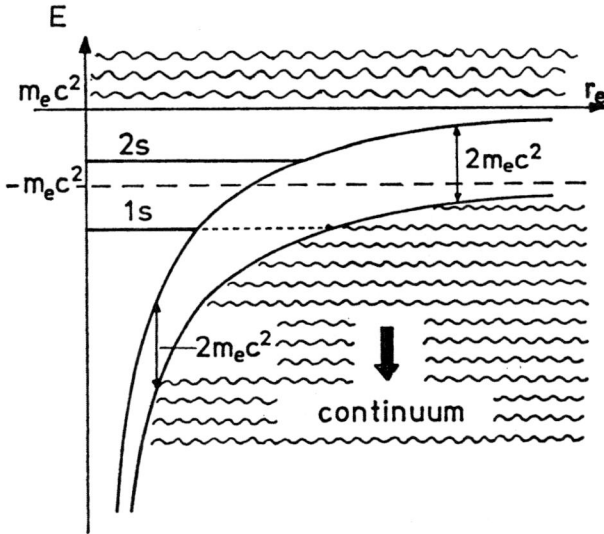

Fig. 9.4.2 Potentials and levels for electrons and positrons.

(From [MÜ 72a]   ).

$$\Gamma[Z(r)] \equiv \Gamma(r) = [Z'(r)]^2 \gamma \tag{9.4.3}$$

as given above. $Z(r)$ is determined by the monopole approximation to the two-center Coulomb potential of the colliding nuclei. Higher-order corrections due to the separation of two centers can be made but do not change the qualitative features. The rate for the escape of a positron with energy E_{pos} at a distance r between the ions is approximated by the golden rule

$$P(r, E_{pos} = m_e c^2 - E) = \frac{1}{\hbar} \Gamma(r) |a(E,r)|^2 \tag{9.4.4}$$

where $|a(E,r)|^2$ is given by eq. (9.4.2). The number of positrons spontaneously created per second with energy E_{pos} at the distance r is then

$$N(r, E_{pos}) = L(r) P(r, E_{pos}) \tag{9.4.5}$$

where $L(r)$ is the hole probability. Integration over t gives the total probability $W(E_{pos}, \theta)$ of positron emission along a given trajectory $r(t)$ which leads to a specific scattering angle θ. Finally, the differential cross-section becomes

$$\frac{d^2\sigma}{dE_{pos} d\Omega} = L_0 W(E_{pos}, \theta) \left(\frac{d\sigma}{d\Omega}\right)_{Ruth} \tag{9.4.6}$$

where L_0 is the K-hole probability at the diving point r_{cr}. The latter is determined by $Z(r_{cr}) = Z_{cr}$. Fig. 9.4.3 shows the function $W(E_{pos}, \theta)$ for a central collision and the energy distribution of the produced positrons, $d\sigma/dE_{pos}$, where $L_0 = 0.01$ has been assumed. The total cross-section for positron production is then

$$\sigma_{tot} = 0.028 \text{ mb} \quad \text{for} \quad U + U \quad \text{(at 6.7 MeV/nucleon)} ,$$
$$= 0.25 \text{ mb} \quad \text{for} \quad U + Cf \quad \text{(at 7.0 MeV/nucleon)} ,$$
$$= 1 \text{ mb} \quad \text{for} \quad Cf + Cf \quad \text{(at 7.2 MeV/nucleon)} .$$

But $L_0 = 0.01$ is expected [AR 73, MO 73] to be too optimistic. A value of 10^{-4} is much more realistic. Hence, $\sigma_{tot} = 0.28 \ \mu b$ for $U + U$ is obtained which indicates that the observation of spontaneously produced positrons is a difficult experiment. One has also to realize that pair production by Coulomb excitation gives a large disturbing background. Experimentally this process has not yet been observed. We refer to $\begin{bmatrix} \text{BE 75, BR 75, GR 75} \end{bmatrix}$ for a more detailed discussion.

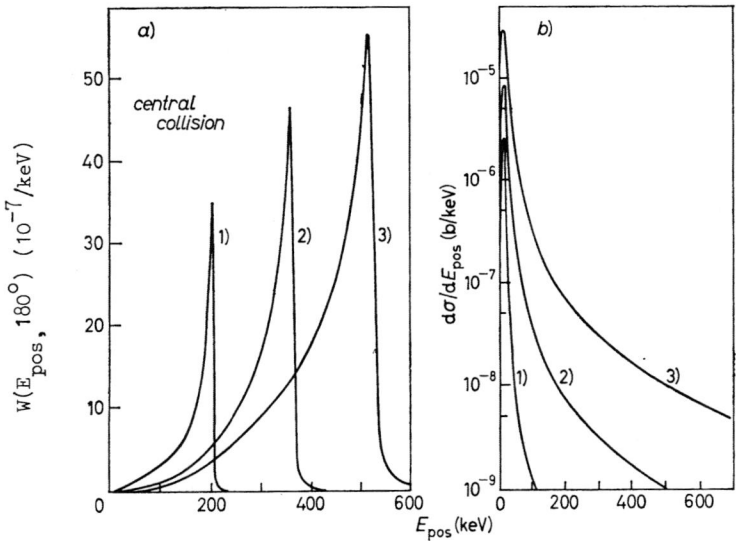

Fig. 9.4.3 The function $W(E_{pos}, \theta)$ and the energy distribution of spontaneously produced positrons for (1) (U (6,7 MeV/amu) + U, (2) U (7.0 MeV/amu) + Cf and (3) Cf (7.2 MeV/amu) + Cf. (From $\begin{bmatrix} \text{PE 73} \end{bmatrix}$).

References

AK 65   A.I. Akhiezer and V.B. Berestetskii, Quantum Electrodynamics (Interscience Publ., New York, 1965) p. 131

AN 75   R. Anholt and T.K. Saylor, preprint (1975)

AR 73   P. Armbruster, private communication, 1973

AR 74   P. Armbruster, G. Kraft, P. Mokler, B. Fricke and H.J. Stein, Physica Scripta $\underline{10A}$ (1974) 175

AR 75   P. Armbruster, private communication, 1975

BA 72   M. Barat and W. Lichten, Phys. Rev. $\underline{A6}$ (1972) 211

BE 63   F. Beck, H. Steinwedel and G. Süssmann, Z. Physik $\underline{171}$ (1963) 189

BE 75   W. Betz, G. Heiligenthal, J. Reinhardt, R. Kent-Smith and W. Greiner, Proc. of the International Conference on the Physics of Electronic and Atomic Collisions, Seattle, 1975

BR 75   J.S. Briggs, Proc. of the International Conference on the Physics of Electonic and Atomic Collisions, Seattle, 1975

CO 34   W.M. Coates, Phys. Rev. $\underline{46}$ (1934) 542

CR 75   Atomic Inner Shell Processes, vol. 1, Ionization and Transition Probabilities, ed. by B. Crasemann (Academic Press, New York - San Francisco - Lund, 1975)

DA 74   C.K. Davis and J.S. Greenberg, Phys.Rev.Lett. $\underline{32}$ (1974) 1215

DI 47   P.A.M. Dirac, The Principles of Quantum Mechanics (Clarendon Press, Oxford, 1947) p.52

FA 61   U. Fano, Phys. Rev. $\underline{124}$ (1961) 1866

FA 65   U. Fano and W. Lichten, Phys.Rev. Lett. $\underline{14}$ (1965) 627

FR 74   B. Fricke and G. Soff, report, GSI-T1-74,(Gesellschaft für Schwerionenforschung (GSI), Darmstadt, 1974)

FR 75   W. Frank, P. Gippner, K.H. Kaun, H. Sodan and Y.P. Tretyakov, Phys.Lett. $\underline{59B}$ (1975) 41

GA 73   J.D. Garcia, R.J. Fortner and T.M.Kavanagh, Rev.Mod.Phys. $\underline{45}$ (1973) 111

GE 69   S.S. Gershtein and Y.B. Zeldovich, Nuovo Cimento Lett. $\underline{1}$ (1969) 835; Sov. Phys. - JETP $\underline{30}$ (1970) 358

GI 74   P. Gippner, K.H. Kaun, H. Sodan, F. Stary, W. Schulze and Y.P. Tretyakov, Nucl.Phys. $\underline{A230}$ (1974) 509 and Phys. Lett. $\underline{52B}$ (1974) 183

GO 74   G.E. Gove, F.C. Jundt and H. Kubo, report UR-NSRL-81 (University of Rochester, 1974)

GR 74   J.S. Greenberg, C.K. Davis and P. Vincent, Phys.Rev.Lett. 33 (1974) 473

GR 75   W. Greiner, Proc. of the EPS Meeting in Bukarest 1975 (to be published)

HU 27   F. Hund, Z.Physik 40 (1927) 742

KE 73   Q.C. Kessel and B. Fastrup, in Case Studies in Atomic Physics III, ed. by E.W. McDaniel and M.R.C. McDowell (North-Holland Publ. Co., Amsterdam, 1973) p.137

LI 67   W. Lichten, Phys.Rev. 164 (1967) 131

LI 74   W. Lichten, in Atomic Physics 4, ed. by G. zu Putlitz et al. (Plenum Publ. Corp., New York, 1974) p. 249

MA 69   C. Mahaux and H.A. Weidenmüller, Shell-Model Approach to Nuclear Reactions (North-Holland Publ. Comp., Amsterdam, 1969)

MA 73   J.R. MacDonalt, M.D. Brown and T. Chiao, Phys.Rev.Lett. 30(1973) 471

MA 75   M.S. Marinov and V.S. Popov, Sov.Phys.- JETP 40 (1975) 621

ME 73   W.E. Meyerhof, T.K. Taylor, S.M. Lazarus, W.A. Little, B.B. Triplett and L.F. Chase, Phys.Rev.Lett. 30 (1973) 1279

ME 74   W.E. Meyerhof, T.K. Taylor, S.M. Lazarus, W.A. Little, B.B. Triplett, L.F. Chase and R. Anholt, Phys.Rev.Lett. 32 (1974) 502

MI 72   A.B. Migdal, A.M. Perelomov and V.S. Popov, Sov.J. Nucl.Phys. 14 (1972) 488

MO 72   P.H. Mokler, H.J. Stein and P. Armbruster, Phys.Rev.Lett. 29 (1972) 827

MO 73   P.H. Mokler, H.J. Stein and P. Armbruster, Proc. of the Int.Conf. on Inner-Shell Ionisation Phenomena, Atlanta, USA, 1972, ed. by R.W. Fink et al. (USAEC, Technical Information Center, Oak Ridge, 1973) vol. II, p.1283

MO 74   P.H. Mokler, S. Hagmann, P. Armbruster, G. Kraft, H.J. Stein, K. Rashid and B. Fricke, in Atomic Physics 4, ed. by G. zu Putlitz et al. (Plenum Publ. Corp., New York, 1974) p.301

MÜ 72a   B. Müller, J. Rafelski and W. Greiner, Z. Physik 257 (1972) 62 and 183

MÜ 72b   B. Müller, H. Peitz, J. Rafelski and W. Greiner, Phys.Rev.Lett. 28 (1972)1235

MÜ 73   B. Müller, J. Rafelski and W. Greiner, Phys.Lett. 47B (1973) 5

MÜ 74   B. Müller, R. Kent-Smith and W. Greiner, Phys.Lett. 49B (1974) 219

PE 72   A.M. Perelomov and V.S. Popov, Sov.Phys.- JETP $\underline{34}$ (1972) 928

PE 73   H. Peitz, B. Müller, J. Rafelski and W. Greiner, Lett. Nuovo Cim. $\underline{8}$(1973) 37

PI 69   W. Pieper and W. Greiner, Z.Physik $\underline{218}$ (1969) 327

PO 45   I.I. Pomeranchuk and I.A. Smorodinskii, J. Phys. (USSR) $\underline{9}$ (1945) 97

PO 70   V.S. Popov, Sov.Phys. - JETP Lett. $\underline{11}$ (1970) 162; Sov.Phys.-JETP $\underline{32}$(1971) 526

PO 71a V.S. Popov, Sov. J. Nucl.Phys. $\underline{12}$ (1971) 235; $\underline{14}$ (1972) 527

PO 71b V.S. Popov, Sov.Phys.- JETP $\underline{33}$ (1971) 665

PO 73   V.S. Popov, Sov.J.Nucl.Phys. $\underline{17}$ (1973) 322

PO 74   V.S. Popov Sov.Phys.-JETP $\underline{38}$ (1974) 18

RA 71   J. Rafelski, L.P. Fulcher and W. Greiner, Phys.Rev.Lett. $\underline{27}$ (1971) 958

RE 69   D. Rein, Z.Physik $\underline{221}$ (1969) 423

SA 72   F.W. Saris, W.F. van der Berg, H. Tawara and R. Laubert, Phys.Rev.Lett. $\underline{28}$ (1972) 717

SA 73   F.W. Saris, I.V. Mitchell, D.C. Santry, J.A. Davis and R. Laubert, Proc. of the Int.Conf. on Inner-Shell Ionisation Phenomena, Atlanta, USA, ed. by R.W. Fink et al. (USAEC, Technical Information Center, Oak Ridge, 1973) vol. II, p. 1255

SM 74   K. Smith, H. Peitz, B. Müller and W. Greiner, preprint (Institut für Theoretische Physik, Universität Frankfurt, 1974)

SO 73   G. Soff, J. Rafelski and W. Greiner, Phys.Rev. $\underline{A7}$ (1973) 903

ST 73   H.J. Stein, P.H. Mokler and P. Armbruster, in Spezielle Probleme der Schwerionenstreuung, Atomphysik und atomphysikalische Experimente an Schwerionenbeschleunigern, GSI 73-11 (Gesellschaft für Schwerionenforschung (GSI), Darmstadt, 1973) p. 106

VO 61   V.V. Voronkov and N.N. Kolesnikov, Sov.Phys. JETP $\underline{12}$ (1961) 136

ZE 72   Y.B. Zeldovich and V.S. Popov, Sov.Phys.-USPEKHI $\underline{14}$ (1972) 673

Subject Index

Selected Issues from

Lecture Notes in Mathematics

SPRINGER TRACTS IN MODERN PHYSICS

Ergebnisse der exakten Naturwissenschaften

Editor: G. Höhler

Associate Editor:
E.A.Niekisch

Editorial Board:
S. Flügge, J. Hamilton,
F. Hund, H. Lehmann,
G. Leibfried, W. Paul

Springer-Verlag
Berlin
Heidelberg
New York

Lecture Notes in Physics